Crops and Livestock Farming

Editor

Dr. Bibek Ghosh

GENE TECH

2017

Gene-Tech Books

New Delhi - 110 002

ISBN 81-89729-79-9

Published by : **GENE-TECH BOOKS**
4762-63/23, Ansari Road, Darya Ganj,
NEW DELHI - 110 002
Phone: 41562849
e-mail: genetechbooks@yahoo.co.in

Printed at : **Tarun Offset Printers**
Delhi

PRINTED IN INDIA

Preface

Farming is the art, science or practice of producing food, feed, fibre and many other desired goods by the systematic raising of plants and animals. Continual improvement in farming methods from pre-history to the present has been the key factor in the extreme specialization of human activity during the historical epoch. About 42 per cent of the world's population is employed in farming, making it by far the most common occupation, yet it accounts for only 4.4 per cent of the gross world product.

Traditional systems of crops and livestock farming are not static. They adapt to changing circumstances such as increased population pressure, use of fertilizers and changing consumption patterns. Farmers, governments, and national and international agencies all have a role in this change, often by keeping track of new technologies and management practices that might be useful for development. The farmers' expertise may combine with the expertise of national and international agencies with access to information from different areas elsewhere in the world. The exchange of experiences between the grassroots/farmer level and national and international level is necessary for the generation and application of appropriate technologies and management techniques that serve to keep farmers in business and to produce enough food for growing populations

This book provides a comprehensive analysis of the modern trends and issues in crops and livestock farming. It will be a valuable reference work on individual farming systems, a rich source of comparative data for researchers, and a guide for advanced courses in farming systems and agriculture.

Editor

Contents

1

Crop Production Systems

Agriculture has always been india's most important economic sector. Since independence in 1947, the share of agriculture in the GDP has declined in comparison to the growth of the industrial and services sectors. However, agriculture still provides the bulk of wage goods required by the nonagricultural sector as well as numerous raw materials for industry. Moreover, the direct share of agricultural and allied sectors in total exports is around 18 percent. When the indirect share of agricultural products in total exports, such as cotton textiles and jute goods, is taken into account, the percentage is much higher.

Dependence on agricultural imports in the early 1960s convinced planners that India's growing population, as well as concerns about national independence, security, and political stability, required self-sufficiency in food production. This perception led to a program of agricultural improvement called the Green Revolution, to a public distribution system, and to price supports for farmers. In the 1980s, despite three years of meager rainfall and a drought in the middle of the decade, India managed to get along with very few food imports because of the growth in food-grain production and the development of a large buffer stock against potential agricultural shortfalls. By the early 1990s, India was self-sufficient in food-grain production.

Agricultural production has kept pace with the food needs of the growing population as the result of increased yields in almost all crops, but especially in cereals. Food grains and pulses account for two-thirds of agricultural production in the mid-1990s. The growth in food-grain production is a result of concentrated efforts to increase all the Green Revolution inputs needed for higher yields: better seed, more fertiliser, improved irrigation, and education of farmers. Although increased irrigation has helped to lessen year-to-year fluctuations in farm production resulting from the vagaries of the monsoons, it has not eliminated those fluctuations.

The agriculture sector, today, provides livelihood to about 64 per cent of the labour force, contributes nearly 26 per cent of Gross Domestic Product and accounts for about 18 per cent share of the total value of the country's exports. It supplies bulk of wage goods required by the non-agricultural sector and raw material for a large section of industry. Per capita net availability of foodgrains went up to a level of 484.1 grams per day in 1998-99 as compared to that of 395 grams in the early fifties. In terms of gross fertiliser consumption, India ranks fourth in the world after USA, the erstwhile USSR and China. The country has the largest area in the world under pulse crops while in the field of cotton, India is the first to evolve a cotton hybrid.

Major Crops

The average rate of output growth since the 1950s has been more than 2.5 percent per year and was greater than 3 percent during the 1980s, compared with less than 1 percent per annum during the period from 1900 to 1950. Most of the growth in aggregate crop output was the result of an increase in yields, rather than an increase in the area under crops. The yield performance of crops has varied widely.

The national growth rates mask variability in the performance of different states, but in the regions with the greatest increases three categories are discernible. The first category includes states or areas that have an exceptionally high agricultural growth rate—Punjab, Haryana, and western Uttar Pradesh. The second is states or areas that have high growth rates, but not as high as the first category—Andhra Pradesh, Maharashtra, and Jammu and Kashmir. A third category has a lesser growth rate and includes Bihar, Gujarat, Karnataka, Orissa, Rajasthan, Tamil Nadu, eastern Uttar Pradesh, and West Bengal. These eight states, however, comprise 55 percent of the total food-grains area.

Some observers believe that the increase in productivity has been an important factor explaining the satisfactory growth of food-grain production since the mid-1960s. However, the gains in productivity remain confined to select areas. Between FY 1960 and FY 1980, yields increased by 125.6 percent in North India.

The increase in the other regions was much less: central India, 36 percent; eastern, 22.7 percent; southern, 58.3 percent; and western India, 31.6 percent. The national average was nearly 40.9 percent. Part of this disparity can be explained by the fact that during this period Punjab and Haryana were way ahead of other states in terms of irrigated area, intensity of irrigation, and intensity of cropping. Availability of irrigation is one of the crucial factors governing regional variations.

As a result of a good monsoon during FY 1990, food grain production reached 176 million tons, 3 percent more than in FY 1989. The production of rice and wheat was 74.6 million and 54.5 million tons, respectively. Among the commercial crops, sugarcane and oilseeds reached production levels of 240.3 million tons and 21.8 million tons, respectively. The increased production in FY 1990 was mainly the result of continuing increases in yields for all the main crops—rice, wheat, pulses, and oilseeds. In the case of oilseeds and sugarcane, higher production was also the result of the increased number of hectares planted.

The growth in food-grain production did not occur in a linear trend, but as a series of spurts depending mostly on the weather, input availability, and price policy. Aggregate growth was composed of an even split between area expansion and yield growth before FY 1964. Since FY 1967, the contribution of growth in yields has become dominant and attests to the vigor with which agriculture has responded to the opportunities opened up by new seed, water, and fertiliser technology.

Food-Grain Production

Food grains include rice, wheat, corn (maize), coarse grains (sorghum and millet), and pulses (beans, dried peas, and lentils). In FY 1990, approximately 127.5 million hectares were sown with food grains, about 75 percent of the total planted area. The total number of hectares increased by 31 percent over the forty-year period from FY 1950 to FY 1990. Most of this increase occurred in the 1950s; there was almost no change in the sown number of hectares through the 1980s. Around 33 percent of cropland was given over to rice, about 29 percent to coarse grains, and the rest evenly divided between wheat and pulses.

Rice, India's preeminent crop, is the staple food of the people of the eastern and southern parts of the country. Production increased from 53.6 million tons in FY 1980 to 74.6 million tons in FY 1990, a 39 percent increase over the decade. By FY 1992, rice production had reached 111 million tons, second in the world only to China with its 182 million tons. Since 1950 the increase has been more than 350 percent. Most of this increase was the result of an increase in yields; the number of hectares increased only 40 percent during this period. Yields increased from 1,336 kilograms per hectare in FY 1980 to 1,751 kilograms per hectare in FY 1990. The per-hectare yield increased more than 262 percent between 1950 and 1992.

Wheat production showed an 843 percent increase, from nearly 6.5 million tons in FY 1950 to 54.5 million tons in FY 1990 to 56.7 million tons in FY 1992. Most of this greater production was the result of an increase in yields that went from 663 kilograms per hectare in FY 1950 to 2,274 kilograms in FY 1990. Along with the excellent performance in yields, improved wheat production resulted from an increase in the area planted from nearly 9.8 million hectares in FY 1950 to 24.0 million hectares in FY 1990.

Sorghum and millet, the principal coarse grains, are dryland crops most frequently grown as staples in central and western India. Corn and barley are staple foods grown mainly near and in the Himalayan region. As the result of increased yields, the production of coarse grains has doubled since 1950; there was hardly any change in the area sown for these grains. The production of pulses did not fare well, increasing by only 68 percent over the four decades. Land devoted to pulses increased by 28 percent, and yields were up by 30 percent. Pulses are an important source of protein in the vegetarian diet; the small improvement in production along with the increase in population meant a reduced availability of pulses per capita.

Before the Green Revolution, coarse grains showed satisfactory rates of growth but afterward lost cultivated areas to wheat and rice, and their growth declined. The area sown with coarse grains increased from FY 1950 to FY 1970 by roughly 20 percent but declined subsequently up to the early 1990s. In FY 1990 the area sown was 3 percent less than in FY 1950 and 20 percent less than in FY 1970. The area sown with two coarse grains, *jowar* (barley) and *bajra* (millet), increased from FY 1950 to FY 1970 and then declined during the 1970s and the 1980s. The area sown with *jowar* increased from 15.6 million hectares in FY 1950 to 17.4 million hectares in FY 1970 and then decreased to 14.5 million hectares in FY 1990. The area sown with *bajra* increased from 9.0 million hectares in FY 1950 to 12.9 million hectares in FY 1970 and stood at 10.4 million hectares in FY 1990. A similar pattern existed for other coarse grains. Overall, India's coarse-grain production increased from 15.4 million tons in 1950 to 29 million tons in 1980 to 33.1 million tons in 1990 and 33.7 million tons in 1993.

Oilseeds

India in the mid-1990s has almost attained self-sufficiency in the production of oilseeds to extract vegetable oil, essential in the Indian diet. Peanuts, grown mainly as a rain-fed crop on part of the semiarid areas of western and southern India, account for the largest source of the nation's production of vegetable oils. The second-ranking source of vegetable oils in the early 1990s was rapeseed. Cottonseed, an important by-product of cotton fiber and once mostly fed to cattle, was another source of vegetable oils. Soybeans and sunflower seeds were relatively new as significant oilseeds, but their production increased rapidly in the 1980s.

The production of oilseeds increased from 5.2 million tons in FY 1950 to 21.8 million tons in FY 1990. Specific information regarding area planted is not available for all oilseeds, but it increased in the 1980s, as did the yields. The growth of production before the mid-1970s was not adequate to meet the needs of the increasing population, and large quantities had to be imported from the 1970s to the mid-1980s, using scarce foreign exchange.

Commercial Crops

India is the largest producer of sugar in the world, harvesting 12 million tons in 1993, followed by Brazil's 9 million tons and China's 7 million tons. Sugar availability per capita increased from 4.7 kilograms per year in FY 1960 to 12.5 kilograms per year in FY 1990, following the more than fourfold increase in production from 57 million tons in FY 1950 to 240 million tons in FY 1990. This increase in production was a result of the doubling of the yield per hectare and a doubling of the area sown with sugar. Imports of sugar were negligible in FY 1992 and FY 1993. However, in the FY 1995 budget presentation to the Lok Sabha in March 1995, Minister of Finance Manmohan Singh said it was necessary to supplement the public distribution system with "necessary imports of sugar."

Raw cotton is the most important nonfood commodity produced on India's farms. Cotton was an important export crop in the 1950s, but thereafter it provided the raw material for India's textile industry, which grew greatly to meet the needs of an expanding population. Cotton fabrics found an expanding international market in the 1980s and earned valuable foreign exchange. The foreign exchange earned from raw cotton, cotton yarn, and fabrics of all textile materials increased from US$163 million in FY 1960 to US$1.4 billion in FY 1980 to nearly US$3.9 billion in FY 1990 and US$3.8 billion by FY 1992. Cotton production increased from 600,000 tons in FY 1950 to nearly 1.7 million tons in FY 1990. These improvements largely resulted from increased yields, as there was little increase in the sown area devoted to cotton.

Raw jute is second only to cotton as a farm-produced industrial raw material. Before partition in 1947, India was the world's main supplier of jute and jute goods used as packaging material. As a result of the partition of India and Pakistan, the main jute growing area was in East Pakistan, and the factories manufacturing jute goods were in West Bengal, which remained part of India after partition. Jute also had been India's main source of export earnings.

As a result, there was a concerted effort to increase raw jute production. The area sown with jute increased from 571,000 hectares in FY 1950 to nearly 1.2 million hectares in FY 1985 but then decreased to 692,000 hectares in FY 1988. Yields increased steadily from 1,040 kilograms per hectare in FY 1950 to 1,803 kilograms per hectare in FY 1990. These two factors combined to more than double jute production from 595 million tons in FY 1950 to 1.4 billion tons in FY 1990, with a maximum production of nearly 2 billion tons in FY 1985. Because technological changes in packaging reduced the worldwide demand for jute, production in the early 1990s was mainly for the domestic market. In FY 1990, jute provided less than 1 percent of export earnings.

Harvesting Season

There are three main crop seasons, namely, *kharif, rabi* and summer. Major *kharif* crops are rice, *jowar, bajra,* maize, cotton, sugarcane, sesame and groundnut. Major *rabi* crops are wheat, *jowar*, barley, gram, linseed, rapeseed and mustard. Rice, maize and groundnut are grown in summer season also.

Land Utilisation

Land utilisation statistics are available for 92.7 per cent of total geographical area of 3,287.3 lakh hectares. According to land use statistics available from States, area under forests had increased from 404.8 lakh hectares in 1950-51 to 688.3 lakh hectares in 1995-96. Net sown area increased from 1,187.5 lakh to 1,422.2 lakh hectares during the same period. Broad cropping pattern indicates that though foodgrains have a preponderance in gross cropped area as compared to non-foodgrains, their relative share came down from 76.7 per cent during 1950-51 to 65.8 per cent during 1995-96.

Seeds

Seed is a critical and basic input for attaining agricultural production and productivity in different agro-climatic regions. The role of quality seed programme, came into prominence about three decades ago. The country made a small beginning with a few thousand quintals of improved seed in the early 60s. Indian seed programme largely adheres to the limited generation system for seed multiplication. The system recognises three generations, namely, breeder, foundation and certified seeds and provides adequate safeguards for quality assurance in the seed multiplication chain to maintain the purity of variety as it flows from the breeders to the farmers.

The level of certified/quality seed distributed to the farmers during 1993-94 to 1998-99 and target for 1999-2000 is given below:

Year	Certified/quality seed distribution (in lakh quintals)
1993-94	62.20
1994-95	65.86
1995-96	69.90
1996-97	73.27
1997-98	75.60
1998-99	83.00
1999-2000	91.00

Seed Industry

Indian seed programme includes the participation of Central and State governments, ICAR, SAU system, public sector, co-operative sector and private sector institutions. Seed sector in India consists of two national level corporations, *i.e.*, National Seeds Corporation (NSC) and State Farm Corporation of India (SFCI), 13 State Seed Corporations (SSCs) and about 100 major private sector seed companies. For quality control and certification, there are 20 State Seed Certification Agencies (SSCAs) and 100 State Seed Testing Laboratories (SSTLs). Though the private sector has started to play a significant role in the production and distribution of seeds particularly after the introduction of the New Seed Policy of 1988, the organised seed sector particularly for food crops and cereals continues to be dominated by the public sector.

The Seeds Act, 1966 provides for the legislative framework for regulation of quality of seeds sold in the country. The Act also provides for a system of certification of seeds on a voluntary basis. The Act further provides for notification of varieties which is a pre-requisite for certification. The Central Seed Committee and its various sub-committees and the Central Seed Certification Board are apex agencies set up under the

Act to deal with all matters relating to administration of the Act and quality control of seeds. In order to ensure equitable distribution of quality seeds to farmers, seed has been declared as an essential commodity under the Essential Commodities Act, 1955. Government of India has promulgated the Seeds (Control) Order, 1988 which seeks to control and regulate seed production and distribution.

In order to encourage export of seeds in the interest of farmers, the procedure for export of seeds has been simplified. Privately developed varieties/hybrids will be allowed to be exported freely except where there is any emergency caused by the natural calamities and seed is required for meeting demand in the country. Export against orders for which production of seed is organised, is allowed without restriction. In other cases, quantitative ceilings have been prescribed up to which export is permitted freely subject to the provisions of the EXIM Policy.

Under National Seeds Project-III (NSP-III), assistance was provided to the SSCs, SSCAs as well as SAUs/ICAR for development of infrastructure to ensure adequate availability of quality seeds to the farmers and bring overall improvement in seed quality control programme. Up to 21 September 1998, 2,662 varieties of different crops have been notified for the purpose of the Seeds Act and 49 varieties were de-notified for the purpose of the Seeds Act, as they were very old and obsolete.

Fertilisers

Fertiliser is a key input for increasing agricultural production. The consumption of chemical fertilisers during 1998-99 is estimated to be 16.7 million tonnes compared to 16.18 million tonnes in 1997-98.

To cushion the impact of decontrol of phosphatic and potassic fertilisers, a concession of Rs 1,000 per tonne on DAP and MOP and proportionate concession on complexes was provided during 1992-93 under a special scheme of concession on sale of decontrolled phosphatic and potassic fertilisers.

Decontrol and consequent price rise had resulted in reduced consumption of phosphatic and potassic fertilisers and deterioration in the NPK use ratio to 9.5:3.2:1 in 1992-93. With the enhancement of these concessions and removal of some other constraints, consumption of phosphatic and potassic fertilisers has increased significantly. As a result, the NPK ratio has improved to 7.9:2.8:1 during 1997-98. Funds amounting to Rs 3,789.94 crore were utilised during1998-99 on the scheme of concession. The Government has decided to keep the MRPs of DAP, MOP and complexes during 1999-2000 at the same level as they were during 1997-98 to 1998-99.

The Government of India is implementing a Centrally-sponsored scheme on balanced and integrated use of fertilisers to popularise the use of organic sources of nutrients through the use of compost, green manures and bio-fertilisers, etc. Under this scheme during the Eighth Plan assistance was given for strengthening of 147 input testing laboratories, *viz.*, fertilisers, seed, soil and bio-fertilisers and setting up of 21 mechanical compost units is various States. In addition, assistance was also given for organising 25,540 demonstrations and training on micro-nutrients and 1,160 trainings on biocomposting under the scheme. About 400 front-line demonstrations were organised on soil test-based crop response to fertiliser application through state soil testing laboratories in collaboration with ICAR. Funds amounting to Rs 17.28 crore were released during the Eighth Plan to various

State governments for implementing the above components of the scheme. The scheme has been approved for implementation during Ninth Plan with budget allocation of Rs 46.11 crore. The scheme envisages setting up of compost plants for converting city waste/ garbage into manure. During 1997- 98 and 1998-99, funds amounting to Rs 355.55 lakh and Rs 159 lakh respectively were released under the scheme.

The Government of India is implementing a Central-sector scheme 'National Project on Development and Use of Biofertilisers'. Under the scheme, a National Biofertiliser Development Centre has been established at Ghaziabad with its six regional centres at Hissar, Jabalpur, Bhubaneswar, Bangalore, Nagpur and Imphal. At present, these centres are producing and distributing more than 375 tonnes of biofertilisers per year. During 1998-99, these centres produced 310 tonnes of bio-fertilisers. Besides, 14 orientation courses, 42 extension workers training, six refresher courses, and 70 field demonstration trials/farmers fairs were organised.

The scheme envisages providing financial support to State government/ government undertakings as well as private entrepreneurs in terms of onetime non-recurring grant-in-aid of Rs 20 lakh for setting up/strengthening of biofertiliser production unit with capacity of 150 tonnes per annum and Rs 1.5 lakh for strengthening of BGA sub-centre.

During the year 1998-99, five bio-fertilisers production units of State government/institutional agencies/ private entrepreneurs have been set up through release of Rs 50 lakh grantin- aid as first instalment. The Government has so far financed 54 bio-fertiliser production units with the central financial assistance and thus created 7,075 tonnes annual bio-fertilisers production capacity in the country.

During the Ninth Plan, the scheme is being implemented with a total out lay of Rs 19.30 crore.

The Government has issued Fertiliser Control Order, 1985 under the Essential Commodities Act, 1955. This Order contains specifications of fertilisers being sold in the country, methods of sampling and analysis as well as provision for appointment of enforcement agencies for regulating the trade and distribution of fertilisers. There are 61 fertiliser control laboratories in various states with a total analysing capacity of 1,06,475 samples per annum. The Government has been implementing a Centralsector scheme on strengthening of Central Fertiliser Quality Control and Training Institute (CFQC & TI) since the Fourth Plan.

During the Ninth Plan, the scheme is being implemented with an outlay of Rs 6.80 crore. The institute is organising training programmes for State Enforcement Officers, Fertiliser Analysts and for foreign trainees from developing countries. During the Eighth Five Year Plan, 34,662 fertiliser samples were analysed and 61 training programmes for Fertiliser Inspectors and Analysts, 178 orientation courses and 105 dealers trainings were conducted by CFQC&TI, Faridabad and its regional lands. Training courses for fertiliser dealers are also being organised in collaboration with the State governments and fertiliser industry. During 1999-2000, the scheme is being operated at the cost of Rs 187 lakh.

Soil and Water Conservation

Soil and Water Conservation measures are one of the essential inputs for increasing agricultural output in the country. These programmes were first launched during the First Plan. From the very beginning emphasis has been on development of technology for problem identification, enactment of appropriate legislation and constitution of policy coordination bodies. While conceptual framework of soil and water conservation activities has not changed, concept of programmes has undergone considerable revision during successive Five Year Plans.

At present the River Valley Project scheme is being implemented in 34 catchments having total treatable area of 18.02 million hectares out of total catchment area of 71.95 million hectares. Since inception up to the end of the Eighth Plan an area of 3.54 million hectares has been treated with various soil and water conservation measures. Similarly, Flood Prone River Scheme is being implemented in 12 catchments having total catchment area of 24.08 million hectares and treatable area of 7.56 million hectares out of which 0.96 million hectares has been treated till the end of 1997-98.

A Centrally-sponsored scheme of reclaimation of alkali soils was taken up in Punjab, Haryana and Uttar Pradesh during the Seventh Five Year Plan. The scheme was further extended to the States of Gujarat, Madhya Pradesh and Rajasthan during the Eighth Five Year Plan. The scheme aims at improving physical conditions and productivity status of alkali soils for restoring optimum crop production. The major components of the scheme include assured irrigation water on farm development works like land levelling, bunding and ploughing, community drainage system, application of soil amendment, organic manures, etc. An area of 0.53 million hectares out of 3.58 million hectares of alkali land has been reclaimed till the end of 1998-99 in the country.

Another project for alkali land reclaimation and development has been taken up in U.P. and Bihar with the help of EEC at an estimated cost of Rs 85.80 crore out of which Government of India's share is Rs 6.88 crore. The project came into existence during 1993-94 and is likely to be completed by 2001. The project envisages reclaiming15,000 hectares of akali soils during its life span of seven years. An area of 18,836 hectares has been reclaimed till the end of 1998-99.

The scheme of Watershed Development Project in Shifting Cultivation Areas (WDPSCA) was launched in seven north-eastern States with an outlay of Rs 45 crore during the Eighth Plan from 1994-95 with 100 per cent Central assistance to the State plan. The scheme aims at overall development of *jhum* areas on watershed basis. The scheme picked up momentum from 1996-97 onwards. During the Eighth Plan an amount of Rs 40.826 crore was released to the states of north-east region for treating 0.81 lakh hectares area through treatment packages. Up to March 1997 an area of 0.71 lakh hectares has been treated with an expenditure of Rs 31.737 crore. During the Ninth Plan, it has been decided to implement the programme under revised guidelines keeping in view the specific terrain, hydrologic and socioeconomic conditions of the areas which are under formulation.

Farm Mechanisation

The Governmental programmes have been oriented to promote farm mechanisation and make the improved and modern agricultural machines available to the farmers throughout the country with a view to increasing the agricultural production, minimising the drudgery associated with farm operations and removing socio-economic disparity among the farmers. Farmers have been provided assistance for owning agricultural machinery including tractors. The infrastructure for human resource development on farm machinery has been expanded with a view to imparting training in the proper selection, operation, repair/maintenance and management of farm machinery. Besides this, farm machines are tested with a view to evaluating their performance characteristics and upgrading their quality.

However, despite a progressive increase in the use of agricultural machines, its sustainable benefits towards agricultural development have not been uniform in as much as the improvement has mostly been in the northern states and in a few pockets/areas in other states where irrigation facilities have been developed.

Although animal and human power continue to be the main power sources in agriculture, yet the power-driven machines, namely, tractors, power-tillers and combine harvesters, continue to be in the frontline of farm mechanisation. Fuel-efficient tractors and better-quality farm machines are now available in the country. The Government efforts have found expression in a substantial increase in the use of agricultural machines by the farmers. The sale of tractors and power-tillers touched an all-time high during 1998-99, *viz.*, about 2,62,322 tractors and 14,488 power tillers. The same holds good for the other agricultural equipments as well. As a result the power now available at the farms is estimated at 1.16 kw/ha during 1998 as against about only 0.27 kw/ha in the early '70s.

During the Ninth Plan, thrust has been given to popularise the improved animal/power driven implements and small tractors. Water management has also been given specific attention and the water saving devices, *viz.*, sprinkler and drip irrigation system would be promoted through oilseeds, pulses, sugarcane and horticultural development programmes. Industrial designs of prototypes and their jigs and fixtures will be developed for arranging manufacture of new and improved agricultural implements. These schemes would give a fillip to the promotion of agricultural mechanisation in the country.

Farm Machinery Training And Testing Institutes

Farm Machinery Training and Testing Institutes set up at Budni (Madhya Pradesh), Hissar (Haryana), Garladinne (Andhra Pradesh) and Biswanath Chariali (Assam) have the capacity of training about 3,500 personnel on various aspects of farm mechanisation and testing of about 65 machines annually. Since inception about 56,500 personnel have been trained and about 1,400 machines tested by these institutes. During 1998-99, these institutes have trained 3,694 personnel and tested 71 machines.

Among the farm machinery training and testing institutes, Budni Institute has the distinction of being accredited by the Organisation for Economic Corporation and Development (OECD) for official testing of tractors as per OECD test code. This institute is also a designated authority for testing of agricultural tractors and allied machines and issuance of certificates for their conformity to Central Motor Vehicle (Amendment) Rules, 1993.

Promotion Of Agricultural Mechanisation Among Small Farmers

Under this Centrally-sponsored scheme, subsidy at the rate of 30 per cent limited to Rs 30,000 is given to the farmers, their groups, registered cooperative societies, agricultural credit societies, multi purpose agricultural farming societies for the purchase of tractors up to 30 power-take-off horse power (PTO HP) along with matching implements. The scheme had been introduced during 1992-93 with the objective of promoting/ popularising farm mechanisation among farmers for increasing the productivity and alleviating the drudgery associated with various farm operations.

During the Eighth Five Year Plan and during the year 1997-98 and 1998-99 funds to the tune of Rs 3,772.80 lakh and Rs 1,669.30 lakh, and Rs 1,636.94 lakh respectively, were released to the States/UTs, under the scheme, for subsidising 12,576, 5,564 and 5,456 tractors. A provision of Rs 2,000 lakh has been made under the scheme during 1999-2000 for subsidising 6,667 tractors.The Department has decided to launch two new schemes during the Ninth Plan, *viz.*, (a) Promotion of Agricultural Equipment in the northeastern States; and (b) conducting studies and formulating long term mechanisation strategies for each agro-economic zone. Provision of Rs 100 lakh and Rs 50 lakh respectively has been made for these schemes during 1999-2000.

State Agro Industries Corporations

State Agro Industries Corporations (SAICs) have been set up to act as catalyst for providing to the farmers access to the various industrial inputs for agriculture. SAICs have since expanded their basic function by commencing the manufacture and marketing of agricultural inputs, implements and machines, after-sales service and promotion and development of agro-based industries. Seventeen SAICs were set up with equity participation of Government of India. In addition, Government of Manipur and Pondicherry set up their own SAICs without equity participation of Government of India. During 1996-97, the combined turnover of all the 17 SAICs was Rs 1,813.76 crore as compared to Rs 1,567.99 crore in 1995-96.

Plant Protection

Plant protection strategy has been reoriented and Integrated Pest Management (IPM) has been adopted as cardinal principle and main plank of plant protection since Seventh Plan. IPM is an eco-friendly approach aimed at minimal use of chemical pesticides by employing available alternative methods for pest control like cultural, mechanical, biological and use of biopesticides.

Emphasis on human resource development is one of the main features of this programme. To achieve this, the Government of India have organised 518 IPM demonstrations in rice, cotton, vegetables, oilseeds and pulses through establishment of Farmers' Field Schools (FFSs) for training and 1,776 agricultural extension officers and 15,564 farmers in the various techniques of IPM. IPM package of practices for five *kharif* and 14 *rabi* crops has been developed by SAUs as per local needs and its adoption.

Plant protection quarantine activities are intended to prevent introduction of exotic pests and diseases into the country as well as control/containment of exotic pests and diseases which have already been introduced by adoption of suitable domestic quarantine measures. This is achieved by implementation of the provisions of Destructive Insects and Pests Act, 1914 (DIP Act, 1914) and the Plants, Fruits and Seeds (Regulation of Import into India) Order, 1989, issue of Phyto-sanitary Certificates for exportable planting material and Post Entry Quarantine of imported plant and planting material. The National Plant Protection Training Institute, Hyderabad conducted 39 long/short-term plant protection courses to impart training to 866 personnel. Continued liaison is maintained with locust-prone countries and the FAO to keep watch on invasion of locusts.

Dryland/Rainfed Farming

Out of about 142 million hectares of net sown area in the country, 92.6 million hectares is rainfed. Crops grown and cropping practices followed in these areas entirely depend on the rainfall which is often erratic and unpredictable. The bulk of the crops like rice, *jawar, bajra,* other millets, pulses, oilseeds and cotton are grown in this area under rainfed conditions. Farmers in these areas, particularly, small and marginal, are still practising subsistence farming and are in the vortex of a vicious circle. Wide fluctuations in production occur in these areas year after year. Increasing and stabilising agricultural production in these areas is, therefore, of crucial importance.

Government has given high priority to the development of dryland areas and, therefore, several programmes and projects have been launched for utilisation of potential of these areas for :

(i) realising the projected requirement of about 240 MT of annual food production by 2000 AD and to smoothen out fluctuation in annual production;

(ii) reducing regional disparities between irrigated and vast rainfed areas;

(iii) restoring ecological balance by greening rainfed areas through appropriate mixture of trees, shrubs and grasses; and

(iv) generating employment for rural masses and reducing large-scale migration from rural areas to already congested cities and towns.

Holistic approach for integrated farming system development on watershed basis in rainfed areas is the main pursuit of the development activities under the NationalWatershed Development Project for Rainfed Areas (NWDPRA) in the Eighth Plan.

National Watershed Development Project For Rainfed Areas

National Watershed Development Project for Rainfed Areas (NWDPRA) was launched in 1990-91 and covers 25

States and two Union Territories. The objective of the project is restoration of ecological balance in rainfed areas and sustainable bio-mass production. It focuses on :

(i) Conservation, upgradation and utilisation of natural endowments in integrated manner with low cost replicable technology and

(ii) generating employment opportunities for the poverty stricken rural masses in the rainfed areas through directly involving the farmers and watershed beneficiaries in planning and execution of all project works in the watershed by developing Self-Help-Groups of *Mitra Krishak Mandals.*

The farming systems approach on watershed management principles has been adopted in order to conserve rainwater and top soils with people's participation. Animal husbandry, agroforestry, horticulture, fisheries development, sericulture, agro-processing, etc., are some of the diverse activities which are also being taken up.

Under this project the target is of treating an area of 28 lakh hectares at a cost of Rs 1,100 crore during Eighth Plan. Taking into consideration the project criterion of covering all the blocks where arable areas under assured means of irrigation is less than 30 per cent, 2,621 blocks have been identified in which a micro-watershed of 500 to 5,000 hectares has been taken up for treatment. In all 2,554 mico-watersheds with an area of 45.84 lakh hectares have been sanctioned by the State governments at an estimated cost of Rs 1,240.79 crore. By the end of the Eighth Plan 43 lakh hectares have been treated throughout the country with an expenditure of Rs 966.93 crore. An amount of Rs 1,030 crore has been provided for treating 2.25 million hectares during the Ninth Plan. Rs 224.36 crore were released for implementation of

NWDPRA during 1998-99 while Rs 227 crore have been allocated for 1999-2000.

World Bank Aided Watershed Development Projects

Integrated Watershed Development Projects (IWDP) for development of plain and hilly areas in seven selected States were launched in 1990-91 with the World Bank funding. The objectives of these projects are minimising ecological degradation and stabilising watersheds through a number of programmes. Measures adopted are treatment of arable, non-arable lands and drainage lines and promotion of sustainable models of production systems.

The project was conceived for integrated development of plain areas on selected watershed basis in Gujarat, Rajasthan and Orissa under IWDP (plains) and similar development of hilly areas, especially of Shiwalik ranges in Jammu and Kashmir, Himachal Pradesh, Punjab and Haryana through IWDP (hills). The projects have been successfully completed on 31 March 1999. The Phase-II of IWDP (hills) has been launched during 1998-99.

Agricultural Development Projects

World Bank assisted Agricultural Development Projects are at present being implemented in the States of Rajasthan, Tamil Nadu, Assam and Bihar for creation of basic infrastructure for multisectoral activities keeping in view the overall agricultural strategies and priorities of beneficiary States.

Uttar Pradesh Diversified Agricultural Support Project

This five-year project (1998-2004) aided by the World Bank with an estimated US $ 160.5 million is being implemented in about 200 selected blocks of 70 districts of Uttar Pradesh. The main project activities include technology generation, technology dissemination, private development and rural infrastructure.

DANIDA Aided Projects

DANIDA aided Watershed Development Project in Karnataka was initially launched in 1989-90. Keeping in view the successful implementation of this project, DANIDA has agreed to the second project to cover an area of 0.45 lakh hectares at a cost of Rs 21 crore. Comprehensive Watershed Development Project (CWDP), Tirunelveli, Tamil Nadu was started in 1991-92. With its successful implementation, the second phase was started in 1995-96 to cover project area of 0.42 lakh hectares with an outlay of Rs 41.70 crore. Another DANIDA aided Project, CWDP, Ramanathapuram was launched in 1994-95 to cover an area of 0.11 lakh hectares at an estimated cost of Rs 13 crore.

The fourth CWDP in Koraput, Orissa was launched in 1993-94 to cover an area of 0.43 lakh hectares at a cost of Rs 13.25 crore. Another CWDP is in operation in Madhya Pradesh since 1997 to cover project area of 0.34 lakh hectares at a cost of Rs 13.10 crore.

KFW-Assisted Projects

FRG (KFW) assisted watershed development project is under implementation in Karnataka since 1994-95 with an approved cost of Rs 55 crore to cover an area of 53,627 ha. in Hassan and Mandya districts. Indo-German Watershed Development Projects in Maharashtra Phase-I was started in 1992 with an approved cost of Rs 19.73 crore to cover an area of 40,000 ha. It is likely to be completed by December 2000. Another project started in June 1997 with an approved cost of DM 2.5 crore to cover an area of 14,500 ha. in Maharashtra will be completed by 2005.

European Economic Community-Assisted Projects

European Economic Community (EEC) - assisted project, namely, Doon Valley Integrated Watershed Development Project is under implementation since April 1993 for a period of nine years in the hill districts of Uttar Pradesh to mitigate on-going degradation of hilly eco-system. The Project will cover an area of 1.72 lakh hectares at an estimated cost of Rs 82.95 crore. The treated area will be 0.37 lakh hectares. An expenditure of Rs 46.71 crore has been incurred on the project till February 1999. Under arable lands tree plantations and under dryland horticulture, homestead gardens, brushwood/loose boulder check dams will be constructed. Establishment of composite nursery, training of farmers and establishment of *Barani Chetna Kendras* will be continued during 1999-2000.

Swiss Assisted Projects

There are two projects in watershed management sector as listed by Swiss Development Corporation. Firstly, the Indo-Swiss Participative Watershed Development Project, Karnataka (ISPWD-K) is under implementation since December 1995. The total estimated cost of the Project was Rs 20.73 crore with targeted area of 30,000 hectares for development. Up to 1997-98 an area of 15,370 hectares had been treated at a cost of Rs 855.70 lakh. The project has since been completed in March 1999. Another project, namely, People Action for Watershed Development Initiatives (PAWDI) Project has been in operation in Rajasthan from October 1995 and will run up to June 1999. During the first phase of three years (1996-99) about 15,000 hectares area in Alwar and Chittorgarh districts has been taken up. The total cost of the Project is estimated at Rs 1,533.90 lakh. An area of 1,845 hectares has been treated during 1998-99 at a cost of Rs 350.21 lakh.

Crop Production Programme

For sustainable agriculture and increasing overall production of cereals, States and Union Territories are assisted through Centrally-sponsored schemes on

(i) Integrated Cereals Development Programme in Wheat-based Cropping Systems Areas (ICDP-Wheat);

(ii) Integrated Cereals Development Programme in Rice-based Cropping Systems Areas (ICDP - Rice); and

(iii) Integrated Cereals Development Programmes for Coarse Cereals based on Cropping Systems Areas (ICDP - Coarse Cereals).

In addition, in order to popularise new varieties, Seed Mini-kits Programme on Rice, Wheat and Coarse Cereals is also being implemented. Similarly, for the development of commercial crops, the programmes implemented are: Centrally-sponsored Intensive Cotton Development Programme (ICDP), Sustainable Development of Sugarcane-based Cropping Systems and Special Jute Development Programme.

For the development of cereals, cropping system approach has been adopted since 1994-95. The emphasis in the crop-oriented programmes is on field demonstrations and farmers' training for transfer of technologies, popularisation of Integrated Pest Management (IPM) and Integrated Nutrient Management, propagation of high-yielding varieties, supply of farm implements to build resource base of the farmers. Farmers are assisted under various components of the schemes implemented in the concerned State/ UT. These programmes and also the sustained efforts made by our farmers have helped to increase production of cereals, cotton and sugarcane during the Eighth Plan, as compared to the terminal year of the Seventh Plan.

Crop Production

Technology Mission on Oilseeds, Pulses and Maize

The Technology Mission on Oilseeds has achieved a major breakthrough in the production of oilseeds. During 1998-99, another record production of 25.3 million tonnes of oilseeds was achieved over 22.02 million tonnes in 1997-98. The oilseed sector has grown to be a major foreign-exchange earner for the country. During 1998-99 export worth Rs 3,028 crore was made of oilseed, castor oil, groundnut, oilmeals, sesame and Niger seed. For increasing the production of oilseeds, the implementation of Centrallysponsored Oilseeds Production Programme (OPP) is being continued during the Ninth Five Year Plan also with enhanced assistance.

Looking at the success achieved in the oilseeds production, pulses were also brought under the ambit of the Technology Mission in 1990. There has been increase in production of pulses after the inclusion in the Mission. During 1998-99, the production of pulses was estimated to be about 15.2 million tonnes. The implementation of Centrally-sponsored National Pulses Development Project (NPDP) is to be continued during the Ninth Five Year Plan also with enhanced assistance.

Maize has also been included in the Technology Mission in 1995 in view of the versatility of the crop and its importance as a foodgrain, animal feed and an industrial raw material. For increasing the production of maize, Accelerated Maize Development Programme

(AMDP) has been formulated during the Eighth Five Year Plan which is proposed to be implemented during the Ninth Plan.

To augment the availability of edible oil, Oil Palm Development Programme was introduced in the Eighth Plan for which an area of 8 lakh hectares has been found suitable in the 11 selected states, out of which an area of 44,000 hectares has been brought under plantation up to 1998-99.

Horticulture

Horticulture includes a wide range of crops, namely, fruits, vegetables, tuber crops, flowers, medicinal and aromatic plants, mushrooms, plantation crops, spices, etc., being grown in varied agro-climatic conditions of temperate, subtropical and arid zones.

Horticulture development in the country continues to make steady progress in the production and export of agricultural products. India has emerged as one of the largest producer of fruits and second largest producer of vegetables after China. India has first position in production of mango, coconut, cashewnut, spices, etc. India is the largest exporter of cashewnut and provides about 40 per cent of the world's total production. India's share in the world trade in spices is around 18 per cent. India is the largest producer of ginger, turmeric and accounts for 65 per cent and 76 per cent respectively of the total world's production.

The area under coconut in 1997-98 was 18.98 lakh hectares and the production, 13,096 million nuts. With this, India has emerged as the largest producer of coconut in the world. The crop sustains nearly 10 million families, contributes 5-6 per cent of the total vegetable oil pool in the country and earns a foreign exchange of Rs 200 crore annually through the exports of coir and coir products. About 90 per cent of the coconut production is accounted for by the states of Kerala, Tamil Nadu, Andhra Pradesh and Karnataka. Coconut cultivation is catching up in non-traditional areas like the north-eastern States, Madhya Pradesh, Bihar, etc.

The area under cashew was 7.5 lakh hectares and production 3.5 lakh tonnes in 1998-99. Kerala, Karnataka, Tamil Nadu and Andhra Pradesh are the major cashewnut producing states. India is the world's leading producer and exporter of cashew accounting for nearly 50 per cent of the world's demand.

The role of horticulture has been fully recognised for having high potential for diversification of hotriculture, better land use with improved productivity and higher income per unit area for the farmers. Besides, it provides nutritional food to millions of people and better employment opportunities. For the year 1999-2000, an allocation of Rs 260 crore has been provided for various Central sector/Centrally-sponsored horticulture development programmes.

2

Animal Husbandry

A large number of farmers depend on livestock for their livelihood. In addition to supplying milk, meat, eggs, and hides, animals, mainly bullocks, are the major source of power for both farmers and drayers. Thus, animal husbandry plays an important role in the rural economy. The gross value of output from this sector was Rs. 358 billion in FY 1989, an amount that constituted about 25 percent of the total agricultural output of Rs. 1.4 trillion.

In FY 1992, India had approximately 25 percent of the world's cattle, with a collective herd of 193 million head. India also had 110 million goats, 75 million water buffalo, 44 million sheep, and 10 million pigs. Milk production in FY 1990 was estimated to have reached 53.5 million tons, and egg production had reached a level of 23.3 billion eggs.

Dairy farming provided supplementary employment and an additional source of income to many small and marginal farmers. The National Dairy Development Board was established in 1965 under the auspices of Operation Flood at Anand, in Gujarat, to promote, plan, and organise dairy development through cooperatives; to provide consultations; and to set up dairy plants, which were then turned over to the cooperatives. There were more than 63,000 Anand-style dairy cooperativè societies with some 7.5 million members in the early 1990s. The milk produced and sold by these farmers brought Rs320 million a day, or more than Rs10 trillion a year. The increase in milk production permitted India to end imports of powdered milk and milk-related products. In addition, 30,000 tons of powdered milk were exported annually to neighboring countries.

Operation Flood, the world's largest integrated dairy development program, attempted to establish linkages between rural milk producers and urban consumers by organising farmer-owned and -managed dairy cooperative societies. In the early 1990s, the program was in its third phase and was receiving financial assistance from the World Bank and commodity assistance from the European Economic Community. At that time, India had more than 64,000 dairy cooperative societies, with close to 7.7 million members. These cooperatives established a daily processing capacity of 15.5 million liters of whole milk and 727 tons of milk powder.

Animal Husbandry and dairy development plays a prominent role in the rural economy in supplementing the income of rural households, particularly, the landless and small and marginal farmers. It also provides subsidiary occupation in semi-urban areas and more so for people living in hilly, tribal and drought-prone areas where crop output may not sustain the family. Animal husbandry output constitutes over 25 per cent of the country's agricultural output. According to provisional estimates of national accounts, the gross value of output for livestock sector during 1997-98 was about Rs 1,11,400 crore as against the total value of output of Rs 4,49,500 crore for the agriculture sector. This excludes the contribution of animal draught power.

India is endowed with the largest livestock population in the world. It accounts for 57 per cent of the world's buffalo population, and 15 per cent of the cattle population. According to the Livestock Census (1992) the country has about 20.5 crore cattle and 8.4 crore buffaloes respectively as compared to 15.5 crore and 4.3 crore in 1951.

Cattle and Buffalo Development

India possesses 27 acknowledged indigenous breeds of cattle and seven breeds of buffaloes. India's cattle are well known in the world for their quality of hardness, endurance and resistance to tropical animal diseases especially tick-born and protozoan infection. Various Central and Centrally sponsored schemes are being implemented for genetic improvement of cattle and buffalo, maintenance of important breeds and production of superior pedigreed bulls, frozen semen, etc.

During 1998-99, 815 artificial insemination centres, one sperm stations, six frozen semen banks, four liquid nitrogen systems and two training centres were established/ strengthened in various states with Central assistance. A Central sector scheme for production of proven bulls by progeny testing under field conditions has been in operation in Andhra Pradesh, Kerala, Maharashtra, Orissa and Tamil Nadu. Besides 44 bulls have been put under test mating in the progeny test centres.

The seven Central cattle breeding farms at Suratgarh (Rajasthan), Dhamroad (Gujarat), Alamadhi (Tamil Nadu), Chiplima, Similigude (Orissa), Andeshnagar (Uttar Pradesh) and Hessarghatta (Karnataka), are engaged in scientific cattle and buffalo breeding programmes and progeny testing of selected breeds of indigenous as well as exotic cattle and buffalo.

Poultry Development

The poultry production in the country has made significant progress over the years due to research and development thrust of the Government and organised private sector. The egg production increased to the level of about 28.5 billion during 1997-98 as compared to 10 billion during 1980-81. It is expected to increase to 30 billion during 1998-99. Currently India ranks fifth in egg production in the world.

Central poultry breeding farms at Mumbai, Bhubaneswar, Hessarghatta and Chandigarh are engaged in scientific poultry breeding programme; developed high egg producing hybrid and fast growing broiler strains and are supplying parent stock chicks. Central Duck Breeding Farm at Hessarghatta is catering to requirements of high egg producing *khaki* campbell breeding stock ducklings to various States/Union Territories.

Four random sample poultry performance testing centres at Bangalore, Mumbai, Bhubaneswar and Gurgaon conduct egg-laying and broiler tests and provide useful information to poultry farmers, hatcheries and breeding organisations about performance of various participating layer and broiler stocks in the country (both from public/private sector organisations).

Central Poultry Training Institute at Hessarghatta is imparting short-term practical courses in different disciplines of poultry to officials of State governments/ Union Territory administrations, agricultural universities, research institutions and private-sector organisations.

Regional feed analytical laboratories at Chandigarh, Mumbai and Bhubaneswar provide feed analysis facilities to farmers, public and private-sector organisations. A scheme for Backyard Poultry has been formulated, which seeks to assist average rural households in increasing their income from poultry by the introduction of improved hybrid varieties of chicks which are suited to Backyard Poultry. The scheme is to be implemented shortly.

During 1998-99, 0.15 lakh egg-type and 0.34 lakh meat-type parent chicks were sold at the Central Poultry Breeding farms and 0.66 lakh ducklings were produced at Central Duck Breeding Farm, Hessarghatta..

Sheep Development

It is estimated that about five million households in the country are engaged in the rearing of small ruminants and other allied activities. The major factors for low productivity of small ruminants in India as compared to the advanced countries are poor exploitation of genetic potential of indigenous animals, low absorption of available technology, inadequate resources of feed and fodder, insufficient health cover, inadequate marketing and credit support, etc. The total production of wool in the country during 1998-99 is estimated at 454.5 lakh kg.

A Central Sheep Breeding Farm with exotic breed of sheep has been established at Hissar. It has a breeding programme with Rambouilit breed as well as cross breeding for production of acclimatised exotic/crossbred superior rams. It has distributed 720 exotic/crossbred rams to different states during 1998-99. The Farm also trains officers and shepherds from different states in modern sheep management. Under the National Ram/ Buck production programme and programme for rabbit development, Central assistance is being provided to State governments on 50:50 basis for strengthening sheep/goat/rabbit farms and state wool boards and to assist them in improving genetic potential of small ruminants.

Meat Production/Processing and Export

A Centrally-sponsored scheme "Assistance to States for improvement/ modernisation of abattoirs/ establishment of carcass utilisation centres and primary hide flaying units" is being implemented in the Ninth Plan period.

The objective of the scheme is to provide wholesome and hygienic meat, gainful utilisation of animal byproducts, prevention of environmental pollution and cruelty to animals. During 1998-99, under Modernisation of Slaughter Houses component of the scheme financial assistance of Rs 4.21 crore has been released to projects at Karnal, Vijayawada, Virar, Solarpur and Hyderabad.

Other projects at Aurangabad, Meerut and Agra are also under consideration for providing assistance. Under the component 'Modernisation of Carcass Utilisation Centres', an amount of Rs two crore has been released to the project at Ghaziabad and the project at Hapur is under consideration. Export of meat and meat preparations during 1997-98 was to the tune of Rs 808.35 crore as against export of Rs 2,190 crore of the livestock products.

Piggery Development

There are more than 128 lakh pigs in the country, of which approximately 14.5 per cent are of graded and exotic variety. At present, there are about 120 pig breeding farms in the country run by the State governments/Union Territories. Exotic breeds like Large White Yorkshire, Hampshire and Landrace are maintained at these farms. These farms breed pigs and supply piglets, boars and sows to the farmers for cross-breeding and improvement of *desi* stock.

A Centrally-sponsored scheme "Assistance to States for Integrated Piggery Development" is being implemented for strengthening the pig breeding farms in the States. During the Eighth Plan, an amount of Rs 800 lakh was released for strengthening the pig breeding farms in the States of Mizoram, Manipur, Assam, Arunachal Pradesh, Andhra Pradesh, Karnataka, Kerala, Madhya Pradesh, Punjab, Uttar Pradesh, Orissa, Meghalaya, Tripura, Nagaland, Sikkim, Bihar, Manipur, Rajasthan and Tamil Nadu. Assistance to pig breeding farms under agricultural universities and *Krishi Vigyan Kendras* is also covered under this scheme.

Feed and Fodder Development

For improvement of vast livestock resources through proper scientific methods, availability of nutritious feed and fodder is essential. To make available scientific fodder production technology, seven regional stations have been established in different agro-climatic zones, *i.e.*, at Hissar (Haryana), Kalyani (West Bengal), Gandhinagar (Gujarat), Alamadi (Tamil Nadu), Hyderabad (Andhra Pradesh), Suratgarh (Rajasthan) and Shehama (Jammu and Kashmir). These are also engaged in production and propagation of certified quality fodder seeds. During 1998-99 these stations produced 155 MT of fodder seeds and conducted 1,635 demonstrations of new fodder varieties.

A Central fodder seed production farm at Hessarghatta (Karnataka) produced 92 MT seeds of different varieties of fodder grasses/legumes during 1998-99. A Central minikit demonstration programme of fodder crops is under implementation for popularising high yielding fodder varieties on a large scale. About three lakh minikits were distributed during 1998-99.

There is a Centrally-sponsored scheme, *viz.*, Assistance to States for Feed and Fodder Development which aims at providing assistance to States for strengthening of State Fodder Seed Farms, establishment of fodder banks, enrichment of straw and cellulosic waste, establishment of silvipasture system, grassland development, etc., in foddercrops.

During 1998-99 financial assistance of Rs 100.86 lakh has been provided for strengthening of two farms in Meghalaya, four in Kerala, and one each in Karnataka and Tripura. Rs 59.85 lakh released for establishment of one fodder bank each at Kerala and Karanataka; Rs 3.33 lakh provided for fodder seed production through registered growers to Punjab, Rs 73.96 lakh provided for enrichment of straw and cellulosic waste in Manipur, Kerala, Punjab, Karnataka, Himachal Pradesh and Tripura; Rs 70.80 lakh provided for establishment of silvipasture system in Mizoram, Kerala, Himachal Pradesh and Rs 41.20 lakh provided for Grassland development including grass reserves in Karnataka and Tripura.

Dairy Development

A significant role has been played by cooperatives in stimulating dairy development. The Operation Flood Programme, which was the world's largest integrated dairy development programme has made considerable progress in achieving its outlined objectives. The programme has since completed its III Phase in April

1996. By March 1998 about 77,531 dairy cooperative societies were organised in 170 milksheds involving over 9.8 million farmer members.

Technology Mission on Dairy Development was launched by the Government of India in August 1988 to accelerate the pace of dairy development in the country. The Mission has so far identified 29 need-based research programmes and assigned them to various national research institutes, agricultural universities and National Dairy Development Board (NDDB) to carry out on a time-bound basis. While 26 projects have been completed others are progressing satisfactorily.

As a result of various measures taken by the Government, there has been a significant increase in the level of milk production in the country to the level of 68.3 million tonnes at the end of Eighth Plan (1996-97) as compared to 17 million tonnes in 1950-51. India's milk output during 1997-98 was about 70.5 million tonnes and is expected to reach the level of 73.5 million tonnes during 1998-99. This makes India the largest producer of milk in the world.

Fisheries

Fisheries play an important role in the economy of India. It helps in augmenting food supply, generating employment, raising nutritional level and earning foreign exchange. The Department of Animal Husbandry and Dairying has been undertaking directly and through State governments/UTs various production, input supply and infrastructure development programmes and welfare-oriented schemes besides formulating/initiating appropriate policies to increase production and productivity in the fisheries sector.

Fish production since 1993-94 is shown in the table below:

(Lakh tonnes)

Year	Marine	Inland	Total
1993-94	26.49	19.95	46.44
1994-95	26.92	20.97	47.89
1995-96	27.07	22.42	49.49
1996-97	29.67	23.81	53.48
1997-98	29.50	24.38	53.88

Development of Inland Fisheries

In recognition of the important role of inland fisheries in overall production of fish the Government has been implementing an important programme in inland sector, *viz.,* Development of Freshwater Aquaculture through the Fish Farmers Development Agencies (FFDAs). These agencies provide a package of technical, financial and extension support to fish farmers. A network of 422 FFDAs is functioning now covering all potential districts in the country. Water area brought under intensive fish culture through the efforts of these FFDAs is 4.56 lakh hectares up to 1997-98. The agencies have trained 5.77 lakh fish farmers in improved practices.

Development of Marine Fisheries

Apart from major fishing harbours, *viz.,* Cochin, Chennai, Visakhapatnam, Roychowk and Paradip, 30 minor fishing harbours and 130 fish landing centres have been constructed to provide landing and berthing facilities to fishing crafts. The Government is providing subsidy to poor fishermen for motorising their traditional craft which increases the fishing areas and frequency of operation with consequent increase in catch and earnings of fishermen. About 33,000 traditional crafts were sanctioned for motorisation up to 1997-98. Improved beach landing craft made of plywood, fibre glass and other light materials are also being supplied to groups of fishermen. The Government has also been operating a scheme of re-imbursing the Central excise duty on HSD oil used by fishing vessels below 20 m length to offset the operational cost, incurred by small mechansied fishing boat operators.

Brackishwater Aquaculture

The objective of this scheme is to utilise the country's vast brackishwater area for shrimp culture. So far, an area of about 20,211 hectares has been developed for shrimp culture. Presently, Brackishwater Fish Farmers Development Agencies (BFDA) functioning in the coastal areas of the country, provide a package of technical, financial and extension support to shrimp farmers. Guidelines prepared for sustainable development and management of brackishwater aquaculture have been circulated to all maritime States and UTs and other user agencies. The guidelines incorporate measures for mitigating the adverse impact, if any, of shrimp farming on the coastal ecosystem.

Development of Fishing Harbour

Fishing harbour with facility to handle large fishing vessels have been provided at major ports. A total of 45 minor fishing harbour and 158 fish landing centres have been sanctioned for construction. Of these 30 fishing harbours at minor ports and 140 fish landing centres have been completed till the end of 1998-99.

Welfare Programmes for Traditional Fishermen

There are two important programmes for the welfare of traditional fishermen, namely, (i) Group Accident

3

Crop Diversification

Crop diversification is intended to give a wider choice in the production of a variety of crops in a given area so as to expand production related activities on various crops and also to lessen risk. Crop diversification in India is generally viewed as a shift from traditionally grown less remunerative crops to more remunerative crops. The crop shift (diversification) also takes place due to governmental policies and thrust on some crops over a given time, for example creation of the Technology Mission on Oilseeds (TMO) to give thrust on oilseeds production as a national need for the country's requirement for less dependency on imports. Market infrastructure development and certain other price related supports also induce crop shift. Often low volume high-value crops like spices also aid in crop diversification. Higher profitability and also the resilience/stability in production also induce crop diversification, for example sugar cane replacing rice and wheat.

Crop diversification and also the growing of large number of crops are practiced in rainfed lands to reduce the risk factor of crop failures due to drought or less rains. Crop substitution and shift are also taking place in the areas with distinct soil problems. For example, the

Insurance Scheme for active fishermen and (ii) Development of Model Fishermen Village. Fishermen are insured for Rs 35,000 in case of death or permanent disability and Rs 17,500 in case of partial disability, a single policy has been taken in respect of all the participating States/UTs through FISHCOPFED and the enhanced premium would be Rs 12 per beneficiary. Annual premium is shared equally between the Centre and the States. About 12.6 lakh fishermen have been insured during 1998-99 . Under the programme of Development of Model Fishermen Villages, basic civic amenities such as housing, drinking-water and construction of community hall for fishermen villages are provided. About 25,683 houses have been constructed for fishermen under the scheme up to 1998-99. In addition to the houses, tube wells and community halls have also been provided.

Specialised Institutes

The Central Institute of Fisheries Nautical and Engineering Training, Kochi, aims at making available sufficient number of operators of deep-sea fishing vessels and technicians for shore establishments. It has a unit each at Chennai and Visakhapatnam. The Institute has an intake capacity of about 200 trainees per year. Integrated Fisheries Project, Kochi, envisages processing, popularising and test marketing of unconventional varieties of fish. The Central Institute of Coastal Engineering for Fisheries, Bangalore is engaged in technoeconomic feasibility study for location of fishing harbour sites. Fishery Survey of India is the nodal agency responsible for survey and assessment of fishery resources under the Indian Exclusive Economic Zone. It has seven operational bases at Mumbai, Porbandar, Kochi, Mormugao, Vishakhapatnam, Chennai and Port Blair.

growing of rice in high water table areas replacing oilseeds, pulses and cotton; promotion of soybean in place of sorghum in vertisols (medium and deep black soils) etc.

The share of the agriculture sector in the total GDP has declined rapidly (24.2 percent) after 1980/81 (34.8 percent) and this trend will continue. By 2020, the share of agriculture in the total GDP of the country is likely to be reduced to 15 percent due to faster development of non-agriculture sectors. The agriculture sector at present employs 60 percent of the country's work force. With the development of alternative sources of employment in the rural areas, viz., agro industries, supportive infrastructure, etc., it is hoped that the share of population dependent on agriculture will come down, though not commensurately, by the year 2020. It is hoped that 45-50 percent of the population will be dependent on agriculture by that time.

India's performance during the post-independence period has been a matter of pride and satisfaction. The agricultural sector has left behind the era of shortages and dependence on imports and arrived at a stage of self-sufficiency and occasional surpluses. The Green, White, Yellow and Blue revolutions have been landmarks that have been claimed and recognised the world over. India is now the largest producer of wheat, fruits, cashew nut, milk and tea in the world and second largest producer of vegetables and fruits. India is the largest producer, consumer and exporter of spices in the world and the largest exporter of cashew.

Foodgrains production has increased four-fold since independence, from 51 million tonnes (Mt) during 1950/51 to 203 Mt during 1998/99. The scourge of severe food shortages is now a thing of the past as is the dependence on imports. India's agriculture has passed through four distinct phases of strategy:

a) starting with the intensification of efforts in identified areas, using traditional technology and expansion of area during the pre-Green Revolution period;

b) through a new strategy of use of modern inputs and high yielding varieties in irrigated areas during the late sixties and the seventies;

c) further through a period of greater focus on management of linkages and infrastructure, such as, marketing, trade and institution building; and,

d) to an era of liberalisation and relaxation of controls during the nineties. The journey has been arduous but rewarding. The agriculture sector has been successful over the past five decades in keeping pace with the rising food demand of a growing population.

This sector provides raw materials to the major industries of the country which are largely agro-based like cotton, sugar, etc. It contributes nearly 16 percent of the country's total export.

India has made tremendous progress in the agricultural sector over the last 50 years. From 'hand to mouth' conditions in the early sixties, we have not only become self reliant in foodgrains but have acquired sufficient resilience to tide over the adverse conditions. Wheat production has increased around 10 times and rice production 4 times during this period. These achievements are the result of a policy framework of improving rural infrastructure including irrigation, research, extension, provision of agricultural inputs at reasonable prices, and marketing support through minimum price mechanism.

In spite of the impressive achievements, the Indian agricultural sector continues to face poor infrastructure conditions. Less than 36 percent of the cultivated land is under any assured irrigation system. Farmers on the remaining two thirds of the land are completely dependent on rainfall, which is also greatly characterised by large variations in terms of precipitation both spatially and in time. For a large majority of farmers in different parts of the country gains from application of science and technology in agriculture have yet to be realised. As a result, the productivity levels of many major crops in India do not compare very favourably with the yields obtained in agriculturally advanced countries.

Further, these factors coupled with high illiteracy constrain the farmer's ability to shift to more remunerative cropping patterns in response to market signals. Therefore, their capacity to take advantage of the opportunities presented by liberalisation of trade is limited. The country's agriculture has gained in strength and resilience since independence, although growth in agriculture is highly skewed over regions and crops. However, the agriculture sector in India is now faced with intense internal and external pressures arising from the impact of policies of economic liberalisation. Efficient and effective management of agriculture will be crucial in the years to come for acquiring enduring self-reliance and ensuring sustainable growth with an emphasis on consideration of equity.

Crop Diversification Patterns

With the advent of modern agricultural technology, especially during the period of the Green Revolution in the late sixties and early seventies, there is a continuous surge for diversified agriculture in terms of crops, primarily on economic considerations. The crop pattern changes, however, are the outcome of the interactive effect of many factors which can be broadly categorised into the following five groups:

a) Resource related factors covering irrigation, rainfall and soil fertility.

b) Technology related factors covering not only seed, fertiliser, and water technologies but also those related to marketing, storage and processing.

c) Household related factors covering food and fodder self-sufficiency requirement as well as investment capacity.

d) Price related factors covering output and input prices as well as trade policies and other economic policies that affect these prices either directly or indirectly.

e) Institutional and infrastructure related factors covering farm size and tenancy arrangements, research, extension and marketing systems and government regulatory policies.

Obviously, these factors are not watertight but inter-related. For instance, the adoption of crop technologies is influenced not only by resource related factors but also by institutional and infrastructure factors. Similarly, government policies - both supportive and regulatory in nature - affect both the input and output prices. Likewise, special government programmes also affect area allocation and crop composition.

More importantly, both the economic liberalisation policies as well as the globalisation process are also exerting strong pressures on the area allocation decision of farmers, essentially through their impact on the relative prices of inputs and outputs. Although the factors that influence the area allocation decision of farmers are all important, they obviously differ in terms of the relative importance both across farm groups and resource regions. While factors such as food and fodder self-sufficiency, farm size, and investment constraints are important in influencing the area allocation pattern among smaller farms, larger farmers with an ability to circumvent resources constraints usually go more by economic considerations based on relative crop prices than by other non-economic considerations. Similarly, economic factors play a relatively stronger role in influencing the crop pattern in areas with a better irrigation and infrastructure potential. In such areas, commercialisation and market networks co-evolve to make the farmers more dynamic and highly responsive to economic impulses.

What is most notable is the change in the relative importance of these factors over time. From a very generalised perspective, Indian agriculture is increasingly getting influenced more and more by economic factors. This need not be surprising because irrigation expansion, infrastructure development, penetration of rural markets, development and spread of short duration and drought resistant crop technologies have all contributed to minimising the role of non-economic factors in crop choice of even small farmers.

What is more, the reform initiatives undertaken in the context of the ongoing agricultural liberalisation and globalisation policies are also going to further strengthen the role of price related economic incentives in determining crop composition both at the micro and macro levels. Obviously, such a changing economic environment will also ensure that government price and trade policies will become still more powerful instruments for directing area allocation decisions of farmers, aligning thereby the crop pattern changes in line with the changing demand-supply conditions.

In a condition where agricultural growth results more from productivity improvement than from area expansion, the increasing role that price related economic incentives play in crop choice can also pave the way for the next stage of agricultural evolution where growth originates more and more from value-added production.

Crop Pattern Changes

Turning now to the socio-economic and environmental consequences of crop pattern changes, the Green Revolution technologies have fomented, among other things, an increasing tendency towards crop specialisation and commercialisation of agriculture. While these developments have positive effects on land/labour productivity and net farm income, they have also endangered a number of undesirable side effects like reduced farm employment and crop imbalances.

Although the expansion of commercialised agriculture has fomented new sets of rural non-farm activities and strengthened the rural-urban growth linkages, it has also weakened the traditional inter-sectoral linkages between the crop and livestock sectors. Besides, crop pattern changes also lead to serious environmental consequences that take such forms as groundwater depletion, soil fertility loss and waterlogging and salinity - all of which can reduce the productive capacity and growth potential of agriculture over the long-term. A classical example is the rice-wheat system in Northwestern India replacing traditional crops like pulses, oilseeds and cotton.

The temporal analysis of the changes in crop pattern is observed both at the national and state level. Crop pattern changes at the all India level are evaluated by considering the area share of crops and crop groups at four time points capturing, respectively, the triennium ending (TE) averages of areas in 1966/67, 1976/77, 1986/87 and 1996/97. These time points have been selected so as to capture the major events and stages in the evolution of Indian agriculture which are of direct relevance to the purpose of this study.

The pre-Green Revolution situation is represented by the 1966/67 period. The post-Green Revolution situation is represented, on the other hand, by three sub-periods. These periods have been chosen deliberately to highlight the temporal differences in the factors contributing to agricultural growth. In the first of these three post-Green Revolution periods, i.e., 1966/67 to 1976/77, the impact of the Green Revolution has mainly been on wheat, especially in the irrigation-wise better endowed Northwestern parts of India. But in the second period, i.e., 1976/77 to 1986/87, the new technologies have expanded to other regions and also covered crops such as rice.

During this period, the Southern and Eastern regions showed remarkable improvement in agricultural growth and productivity. More or less during the same period, the benefits of the new crop and irrigation technologies that were confined earlier to large and medium sized farms have also started percolating to the small farms. The third period, i.e. 1986/87 to 1996/97, is characterised by a number of policy changes including the launching of the Technology Mission on Oilseeds (TMO) as well as price support and stabilisation policies for oilseed crops. During this period there is also a marked shift in the pattern of agricultural investment in favour of the private sector. This period not only covers such important policy changes as unleashed by agricultural liberalisation policies but also represents the consolidation phase of rural infrastructure. Since these four periods can capture well the changes in the set of factors that affect crop pattern, they can provide a better analytical framework for evaluating both the nature and technical context of crop pattern shifts observed not only at the national level but also at the state level.

The analysis of crop pattern changes to be attempted at the macro level (national) will focus on three main aspects. These aspects are:

a) the nature and direction of area shifts across crops and crop groups observed through time,

b) the implications of these shifts for crop diversification and balance in the inter-crop allocation of existing and additional areas brought under cultivation, and

c) the output and productivity impact of crop pattern changes.

Temporal Changes

The temporal behaviour of crop pattern changes at the all India level can be seen from Table 1 and Table 2 that show, respectively, the area share of main crop groups and major crops for the four periods. Though obvious, it needs to be stated that the changing area share of crops is due as much to shift in area under other competing or alternative crops as to the relative area allocation of fresh areas brought under cultivation. In any case, the changing area share of crops does capture the ongoing changes in the comparative advantage calculus of farmers. The changes in the comparative advantage of crops reflect, in reality, the ongoing changes in relative prices of inputs and outputs, production conditions (including irrigation expansion), development and spread of new crop and farm technologies, extension and input support policies and trade policies and domestic regulations. As such, the changing area share of crop pattern, though looking deceptively simple, becomes a useful tool for understanding the direction in which crop pattern changes are influenced by the variations in the comparative advantage of crops and crop groups.

Table 1. Temporal Change (percent) in the Area Share of Main Crops and Crop Groups, 1966/67 to 1996/97

No.	Crops	1966/67	1976/77	1986/87	1996/97
1	Rice	23.90	24.13	24.75	25.29
2	Wheat	11.99	10.01	9.62	15.03
3	Coarse Cereals	26.49	29.20	28.09	18.70
4	All Cereals	62.38	63.34	62.47	59.03
5	Gram & Pigeon Pea	7.26	6.48	6.25	6.30
6	Other Pulses	8.06	7.99	7.85	7.45
7	All Pulses	15.32	14.46	14.10	13.75
8	All Foodgrains	77.70	77.80	76.56	72.78
9	Sunflower & Soybean	0.54	0.79	0.91	0.77
10	All oilseeds	11.27	11.24	12.07	15.35
11	Fibres	6.28	5.25	5.18	5.72
12	Spices	0.64	0.83	1.02	1.12
13	Fruits & Vegetables	0.75	0.92	1.15	1.40
14	Other Field Crops	3.04	3.54	3.51	3.23
15	Plantation Crops	0.33	0.42	0.51	0.41
16	All Commercial Crops	11.03	10.96	11.36	11.87
17	All Non-Cereals	37.62	36.66	37.53	40.98

Although cereals gained a marginal increase in area share in the first decade of the Green Revolution, their area and share declined gradually thereafter. Between 1966/67 and 1996/97, 3.35 percent of the gross cultivated area (GCA) - representing approximately about 5.7 million hectares (m/ha) - has shifted from cereal crops to non-cereal crops. Since the area share of pulses taken as a group also declined by 1.57 percent during the same period, the area share of foodgrains as a group declined by 4.92 percent during 1966-97. In area terms, the shift from foodgrains to non-foodgrains involves an approximate area of about 8.36 m/ha. While cereals and pulses have lost area, the major gainers of this area shift are the non-foodgrain crops especially oilseeds. The area share of oilseeds as a group that has gone up by 4.08 percent accounts for about 83 percent of the 8.36 m/ha involved in the area shift between 1966/67 and 1996/97.

Table 2. All India: Temporal Change (percent) in the Area Share of 52 Major Crops, 1966/67 to 1996/97

No.	Crops	1966/67	1976/77	1986/87	1996/97
1	Rice	23.90	24.13	24.75	25.29
2	Wheat	11.99	10.01	9.62	15.03
3	Sorghum	1.82	1.65	0.77	6.75
4	Pearl Millet	8.03	7.00	6.52	5.81
5	Maize	3.23	3.73	3.51	3.60
6	Finger Millets	1.70	1.58	1.44	1.05
7	Small Millets	3.06	2.88	1.87	1.01
8	Barley	8.66	12.37	13.97	0.49
9	Gram	5.55	4.86	4.35	4.27
10	Pigeon Pea	1.71	1.62	1.90	2.04
11	Blackgram	0.00	1.33	1.85	0.62
12	Greengram	0.00	1.50	1.78	0.60
13	Horsegram	0.00	1.26	1.09	0.21
14	Peas & Beans	0.00	0.43	0.29	0.14
15	Lentil	0.00	0.57	0.63	0.22
16	Lathyrus	0.00	0.98	0.71	0.19
17	Mothbeans	0.00	1.21	0.86	0.12
18	Other Pulses	8.06	7.99	7.85	7.45
19	Groundnut	4.99	4.44	4.27	4.55
20	Castor	0.28	0.30	0.38	0.45

21	Sesamum	1.74	1.39	1.30	1.15
22	Rape & Mustard	1.97	2.11	2.34	3.82
23	Linseed	1.17	1.27	0.80	0.52
24	Safflower	0.21	0.42	0.55	0.43
25	Nigerseed	0.33	0.38	0.36	0.34
26	Sunflower	0.00	0.19	0.52	1.20
27	Soybean	0.00	0.06	0.82	1.83
28	Coconut	0.59	0.68	0.73	1.07
29	Cotton	5.39	4.54	4.38	5.11
30	Jute	0.54	0.41	0.56	0.46
31	Mesta	0.23	0.21	0.18	0.11
32	Sunnhemp	0.12	0.09	0.06	0.04
33	Pepper	0.07	0.07	0.07	0.11
34	Chillies	0.46	0.46	0.51	0.52
35	Ginger	0.02	0.02	0.03	0.04
36	Turmeric	0.04	0.05	0.06	0.08
37	Coriander	0.00	0.16	0.23	0.25
38	Cardamom	0.04	0.06	0.07	0.06
39	Garlic	0.00	0.02	0.04	0.06
40	Banana	0.14	0.14	0.18	0.25
41	Potato	0.31	0.38	0.51	0.68
42	Sweet Potato	0.12	0.15	0.11	0.08
43	Tapioca	0.18	0.24	0.17	0.15
44	Onion	0.00	0.00	0.16	0.23
45	Papaya	0.00	0.01	0.02	0.01
46	Sugar cane	1.73	1.77	1.78	2.39
47	Tobacco	0.27	0.25	0.25	0.24
48	Arecanut	0.09	0.11	0.11	0.15
49	Guarseed	0.95	1.41	1.37	0.45
50	Tea	0.23	0.23	0.24	0.25
51	Coffee	0.02	0.08	0.13	0.05
52	Rubber	0.07	0.11	0.13	0.10

As we consider the share of individual crops within cereals, although the share of cereals as a group has declined, the area share of rice has increased continuously over all the four periods. Wheat, although having a declining area share until 1986/87, also gained in its share when the entire period is considered. Thus, the area loss of cereals can be attributed entirely to the declining area share of coarse cereals, especially sorghum, pearl millet, barely and small millets. It can be noted that even within coarse cereals, the area share of maize shows a marginal improvement over the years. Within oilseeds, the crops showing steady improvement in their area share are: rapeseed and mustard, soybean and sunflower.

Among these three oilseeds gaining in area share, rapeseed and mustard are substantially grown as intercrops with wheat. On the other hand, the area shares of other oilseeds including groundnut but excluding coconut, which is more a plantation crop than field crop, have either fluctuated or declined. The area share of groundnut, though improved during the last period, has declined as compared to its share in the pre-Green Revolution period. But, the declining area share of crops - especially those with only a marginal change in their area share - need not necessarily imply a decline in the actual area under these crops.

Since the Gross Cropped Area (GCA) is constantly increasing over time, partly through an expansion of net sown areas as in the initial stages of the Green Revolution and partly through increasing intensity of cropping mainly by irrigation expansion, the declining area share can coincide with an increase in absolute increase in the area under crops. Although the increase in the area share of other commercial crops is not as dramatic as that of oilseeds, it is still notable because of its implications for the direction of Indian agriculture.

But, among these other commercial crops that cover fibres, spices, fruits and vegetables, and other field crops such as tobacco and sugar cane and plantation crops, only spices, fruits and vegetables show a steady improvement in their area shares, whereas others show mostly a declining trend. This is particularly true for fibres and other field crops that have over four fifths of the total area under the broad group of other commercial crops. However, sugar cane, included in the category of other field crops, shows an increase in its area share. This is also true for cotton included in the fibre category. While all spice crops show a gradual increase in their area share, only three of the six crops included in the fruits and vegetables category show a gain in their area share over the years. These crops are banana, potato and onion.

It is clear that most of the area shifts that occurred during the three decades between 1966/67 and 1996/97 is from coarse cereals to oilseeds. Three notable aspects of this area shift towards oilseeds can be noted.

First, most of these area shifts have occurred particularly during the decade ending 1996/97. A protective trade environment, favourable price policy and the connecting of the Technology Mission on Oilseeds (TMO) during this period have all enhanced the comparative advantage of oilseeds. The area share of oilseeds that increased just by a percentage each during the two decades before 1986/87 has risen suddenly by 43 percent during the decade ending 1996/97.

Second, the area shift during this period comes mainly from barley and pulses other than pigeon pea. Since most of those crops losing their area share are usually grown under rainfed conditions where oilseeds can also be grown, the area shift can be said to involve mostly rainfed areas, although comparative advantage and crop rotation considerations often favour oilseeds even in groundwater irrigated areas.

Thirdly, there is also a significant area shift within oilseed crops. For instance, while the area shares of rapeseed and mustard, sunflower and soybean are increasing steadily, those of sesamum, linseed and nigerseed are declining gradually. Thus, the area shift has favoured only a sub-sector within the oilseed sector partly because of constant changes in the comparative advantage of different oilseeds and partly because of the impact of changing consumers' preferences on the relative demand of oilseeds.

Crop Diversification and Composition

Area shifts and crop pattern changes can lead either to crop specialisation or to crop diversification. The area share of foodgrains increased during 1967-76 due partly to their yield advantages created by irrigation expansion and Green Revolution technologies and partly to government policies pursued to encourage food production and eliminate food imports. As a result, there was a tendency towards cereal centered specialisation. But, later when increased productivity of foodgrains, especially cereals, made it possible to allocate more area to other crops such as oilseeds with a severe supply shortage, the specialisation tendency witnessed earlier has given room for overall crop diversification.

Even within such an overall crop diversification, it is also possible to see specialisation tendencies within each crop group. For instance, within cereals, the declining share of coarse cereals and increasing share of wheat and rice indicates a rice and wheat centered specialisation. Such a specialisation has become possible because within the achievement of self-sufficiency, mainly through an expanding production of wheat and rice, the food implication of coarse cereals has declined almost coinciding with their declining demand prompted by an increasingly incurring income level.

On the other hand, within oilseeds, even though groundnut still has a dominant area share, the growth of area under rapeseed and mustard, sunflower and soybean indicates an ongoing structural change leading to diversification within the oilseed sector. The issue of crop diversification, although considered in area terms, assumes significance in view of its effects on the supply-demand balance of main crops and crop groups. The rice and wheat-centered specialisation, for instance, indicates an increase in the supply of wheat and rice but a reduced supply of coarse cereals. Since the demand for coarse cereals is declining and that for wheat and rice is increasing due to changes in the income pattern, the changes in their supply are actually necessary to achieve the required demand-supply balance. A similar line of argument can also be extended to other crops where in addition to domestic demand, international demand and supply also assume significance.

Clearly, during the Green Revolution period, the foodgrains category has evinced an increasing tendency towards specialisation with an unbalanced area composition, whereas the non-foodgrains category has shown an increasing tendency to diversify with an improved balance in their area composition. But, during 1976-97 the non-foodgrains category has also caught up with the specialisation tendency and area imbalance that characterised foodgrains throughout the entire period. The reason for the specialisation tendency within foodgrains is rather obvious in view of the declining area share of coarse cereals and the increasing area share of rice and wheat. The changing area share of crops can also explain the specialisation tendencies observed among the non-foodgrains.

The area shares of oilseeds and other commercial crops have increased especially during the post Green Revolution period characterised by a prices and market-responsive agriculture inevitable in the aftermath of achieving food self-sufficiency. It is important to note, however, that oilseeds are more diversified with an enhanced balance in their area composition. Within foodgrains, on the other hand, while cereals tend towards specialisation with an increasing imbalance in their area composition, pulses evince a tendency to diversify with a declining imbalance in their area composition.

The latter result is mainly due to the fact that the area shares of the dominant pulses such as gram have declined to give room for other pulses such as pigeon pea. But it should be noted that diversification and area balance achieved by pulses have occurred only with a reduced area share. Among the non-foodgrain crops, spices are notable not only for their increasing tendency towards diversification but also for a declining imbalance in their area composition. In contrast, fibres, fruits and vegetables, other field crops and plantation crops have all evinced a specialisation tendency especially after 1986/87.

Productivity Effects

The evaluation of crop pattern changes and their implications for crop diversification and area composition cannot be completed without considering their output and productivity effects. The production and productivity effects of area shifts can be evaluated by considering both the growth rates of output, area, and yield of various crop groups as well as the crop group specific patterns evident in the relative contributions of area and yield to overall output growth. Table 3 shows the growth rates and sources of growth for the major crop groups at the all India level for the three periods, i.e., 1966-76, 1976-86 and

1986-97. The shift of area from cereals has not led to any decline in their output thanks to an impressive performance of their yield levels.

Table 3. Temporal Change in the Sources of Growth of Main Crop Groups

Crops	Percent Growth Rate in Output Growth from					
	Periods	Output	Area	Yield	Area(%)	Yield (%)
Cereals	1766-76	2.43	0.47	1.96	19.28	80.72
	1976-86	2.55	0.14	2.41	5.65	94.35
	1986-97	2.78	-0.25	3.03	-9.14	109.14
Coarse Cereals	1766-76	0.60	1.05	-0.45	175.87	-75.87
	1976-86	0.05	-0.02	0.07	-50.20	150.20
	1986-97	1.14	-4.36	5.51	-381.16	481.16
Pulses	1766-76	0.84	0.43	0.41	51.25	48.75
	1976-86	1.35	0.21	1.15	15.22	84.78
	1986-97	-0 88	0.32	-1.20	36.39	-136.39
Foodgrains	1766-76	2.25	0.46	1.79	20.46	79.54
	1976-86	2.44	0.16	2.29	6.40	93.60
	1986-97	2.53	-0.15	2.67	-5.83	105.83
Nine Oilseeds	1766-76	2.55	0.55	2.01	21.42	78.58
	1976-86	3.02	1.22	1.80	40.31	59.69
	1986-97	6.28	3.05	3.22	48.68	51.32
*R&M+SF+SB	1766-76	4.69	2.29	2.40	48.79	51.21
	1976-86	8.19	4.93	3.27	60.13	39.87
	1986-97	10.92	7.38	3.53	67.62	32.38
Comm. Crops	1766-76	2.95	0.78	2.17	26.49	73.51
	1976-86	1.55	0.37	1.18	23.96	76.04
	1986-97	4.27	2.23	2.04	52.18	47.82
Non-cereals	1766-76	1.87	0.59	1.28	31.59	68.41
	1976-86	2.81	0.57	2.24	20.38	79.62
	1986-97	5.00	1.89	3.12	37.67	62.33
All Crops	1766-76	2.62	0.51	2.11	19.48	80.52
	1976-86	1.94	0.30	1.64	15.63	84.37
	1986-97	3.78	0.60	3.17	15.98	84.02

* R & M = Rapeseed & Mustard; SF = Sunflower; SB = Soybean.

The productivity improvement in cereals has been such that it compensates even for the area loss of cereals occurred during 1986-97. Notably, in the case of coarse cereals that have suffered a heavy area loss ever since 1976, yield improvement has been substantial especially in the last period. As a result their output growth, which was only marginal during 1976-86, has jumped to 1.14 percent during 1986-97. For pulses, on the other hand, even though area growth has increased from 0.21 to 0.32 percent between the last two periods, their output growth turns out to be negative due to a fall in their yield levels.

Foodgrains showing a steady rise in their output growth have a story similar to that of cereals as the contributions of their yield growth is increasing; whereas, those of their area growth are declining even to become negative in the recent period. Turning to the growth rates and growth sources of oilseeds, it is useful to distinguish the trend observed for the three oilseeds, i.e., rapeseed and mustard, sunflower and soybean from that observed for the oilseeds as a whole. This is because of the fact that the area shares of these three oilseeds have grown tremendously since 1976.

Due to their faster expansion, their combined area share of 6.85 percent observed at present is higher than the combined share of both groundnut and sesamum, which dominated the oilseed sector in the pre-Green Revolution era. While the output growth of nine oilseeds has been impressive as compared to that of other crop groups, it is not as impressive as the combined output growth of the three fast growing oilseeds. However, there is a marked difference in the sources of output growth between the nine oilseeds and its sub-group. Although the contribution of area growth dominates that of yield growth in both cases, the contributions of area growth are far higher in the case of the three oilseeds as compared to that in the case of nine oilseeds. It should be noted that it is the higher area growth observed in the case of the three oilseeds, which has also compensated for the declining area share of some of the traditional oilseeds such as sesamum, linseed and nigerseed. Since the yield growth of the three oilseeds is also higher than that obtained for the nine oilseeds, the productivity improvement in the case of former is also relatively higher than that of the traditional oilseeds. There is also a notable shift in the sources of output growth among the commercial crops (i.e., crops other than cereals, pulses, and oilseeds) as yield growth that dominated over the area growth during 1966-1986 becomes dominated in turn by a area growth during 1986-97.

The implications are clear that area shifts from crops need not be a problem as long as their productivity levels are increasing faster to compensate for their declining area share. This is what has happened especially in the case of coarse cereals that have been a net donor of area to oilseeds and other commercial crops. Despite their negative area growth of -4.36 percent, they managed to maintain a positive output growth of 1.14 percent thanks to an impressive yield growth of 5.51 percent.

Since this pattern is more or less repeated in the case of cereals as a whole, it is reasonable to say that the area shift has not affected food security. On the other hand, since the area shifts were from low-value coarse cereals to high-value oilseeds and since such shifts were accompanied by yield improvements within the oilseed sector, the area shifts have not only increased the overall output of edible oils but also contributed to an enhancement of the income level of farmers. In this sense, the area shifts have actually contributed to broaden the foundation of food security.

Crop Diversification as a Strategy for National Commitments

The agricultural growth rate of around 2.7 percent per annum in the post-independence period was much higher than the negligible growth rate of 0.3 percent per annum in the first half of this century. It is not only in the areas of foodgrain production but also the production of commercial crops like cotton, oilseeds, sugar cane, fruits and vegetables as well as livestock products and fisheries have recorded significant increases since independence. Poverty eradication is one of the major objectives of plan development.

The magnitude of the problem is quite staggering. The incidence of poverty declined from 54.9 percent in 1973/74 to 36 percent in 1993/94. The absolute number of population did not decline much throughout this period of 20 years. There were 321 million poor in 1973/74 and 320 million in 1993/94. In the rural areas, the corresponding numbers were 261 million and 244 million. The land resources are limited and the average size of holding in India is only 1.57 ha for the census year of 1990/91. The main determinants of poverty are:

i) lack of income and purchasing power contributable to lack of productive employment;

ii) the continuous increase in price of food, specially foodgrains which account for 70-80 percent of the consumption; and

iii) inadequacy of social infrastructure, affecting the quality of life of the people and their employability.

The National Agenda for Governance of the present government has given top priority to doubling food production in the next 10 years. The food includes foodgrains (rice, wheat, coarse cereals, pulses), oilseeds, sugar (sugar cane), fruits and vegetables, meat, milk, and fish. The Action Plan envisages a detailed strategy and specific problems of productivity to substantially increase the supply of various food items in such a way that the demand for such items for the entire population is comfortably met and some exportable surplus also becomes available. The development strategy to be perused in the medium term has been consciously interwoven with the country's food security concern.

Natural Resource Management

It is known fact that there is little scope for further expansion of the net sown area and that land scarcity will become an acute feature of the rural economy. Water is a precious national asset and there are several concerns regarding water resources in the country. Therefore, a judicious use of land and water resources will have to be the central theme for sustainability of agricultural growth. There has been a growing concern in recent years about the deteriorating conditions of soil health and water resources due to improper management and pollution. The deterioration in land and water resources has been in the form of land degradation, waterlogging and decline in watertable.

There is a greater need to have an integrated approach in the management of plant nutrients, chemicals and taking effective measures to deal with the overall pollution problems. There are several possible technologies and alternatives to reduce the use of chemicals in agriculture. These alternatives are not perfect substitutes to chemicals but adoption of these can substantially reduce the adverse impact on environment. Proper land and water management policies would reduce environmental degradation. Community and village institutions should be encouraged to participate in protecting natural resources from degradation.

Programmes for regeneration of land and water resources will be strengthened.

Agricultural Planning

A new approach to agricultural planning - the Agro Climatic Regional Planning (ACRP) was put into action in 1988. This holistic approach digresses from the sectoral approach of planning practiced so far in the country. It explicitly recognises the local resource endowments and constraints of the agro-climatically homogeneous regions, quite often cutting across the States. The ACRP can be considered as a bridge between the resource base and decentralised planning, which aims at providing a scientific support to planning for attainment of sustainability and having due consideration of basic resources and the local needs.

The project was initiated by regionalising the country into 15 zones/regions and later into 73 sub-regions and subsequently demarcation of sub-zone/region within a State using district as the lowest unit of analysis. The principles used for this sub-regionalisation were those which relate intrinsically with the character of the agricultural economy like soil, climate, rainfall, etc. In a recent study, agro-climatic zones have been delineated into 4 agro-economic zones to address the issues of poverty, productivity and sustainability. These agro-economic zones are:

i) High Productivity Zone (103 districts);

ii) Low Productivity - High Potential Zone (181 districts);

iii) Low Productivity Zone (179 districts); and

iv) Ecologically Fragile Zone (Himalayan and Desert areas).

Constraints in Crop Diversification

Crop diversification in the country is taking the form of increased areas under commercial crops including vegetables and fruits since independence. However, this has gained momentum in the last decade favouring increased area under vegetables and fruits and also to some extent on commercial crops like sugar cane, cotton and oilseeds crops specially soybean. The major problems and constraints in crop diversification are primarily due to the following reasons with varied degrees of influence:

i. Over 117 m/ha (63 percent) of the cropped area in the country is completely dependent on rainfall.

ii. Sub-optimal and over-use of resources like land and water resources, causing a negative impact on the environment and sustainability of agriculture.

iii. Inadequate supply of seeds and plants of improved cultivars.

iv. Fragmentation of land holding less favouring modernisation and mechanisation of agriculture.

v. Poor basic infrastructure like rural roads, power, transport, communications etc.

vi. Inadequate post-harvest technologies and inadequate infrastructure for post-harvest handling of perishable horticultural produce.

vii. Very weak agro-based industry.

viii. Weak research - extension - farmer linkages.

ix. Inadequately trained human resources together with persistent and large scale illiteracy amongst farmers.

x. Host of diseases and pests affecting most crop plants.

xi. Poor database for horticultural crops.

xii. Decreased investments in the agricultural sector over the years.

Globalisation and Crop Diversification

With the advent of WTO and India being a member and signatory to GATT, the scenario of the agricultural sector will not be the same as that of past. With the liberalisation of trade and providing the market access of agricultural produce between the different countries, the country will be required to promote much more diversified agriculture.

For crops on which we have substantial area and production, specially foodgrains, the import market has to be insulated through increased productivity which gives us a kind of comparative advantage and also a level playing field so that large scale importation is contained and farmers interests are protected. The crops which are traditionally exported like basmati rice and spices and condiments also need to be supported in terms of area expansion and quality improvement to look towards much more opportunity for export.

Crop diversification in the areas of certain tropical fruits and also a few vegetables also need support for both production and post-harvest handling in terms of their export opportunity. Accelerated growth in fruits and vegetables production is also required for improved nutrition of the country's population. In future, with improved living standards along with increased purchasing power, more and more people will look for nutritional and quality foods which will also call for greater crop diversification.

There are some production areas such as food crops, plantation crops, poultry, dairy, sugar, cotton and oilseeds in which India has made its mark. There are some in which its emerging strength is already evident - sericulture, marine and inland fisheries for example. There are also others which now attract less attention, but in which the competitive advantages that India possesses can put it on the top of the world. No country grows such a wide range of fruits, vegetables, and flowers and in such abundance as India and yet it has no record worth mentioning in horticultural exports. The rich variety when processed and marketed, can help India take care of the health needs of its population besides being major export commodities.

Emerging Technology

It is being increasingly realised that agriculture is no longer a subsistence activity carried out by peasant cultivators, but rather an enterprise and manufacturer of biomass using land, water, genetic material and the latest in technology. The agriculture of the twenty first century will increasingly be a farmers' entrepreneurship harnessing technologies to optimise returns from his land and investments he makes on it.

Biotechnology and genetic engineering in crops with focus on primary productivity and also on many quality traits will go a long way to improving the yield and quality of many important crop plants. With the advent of such emerging technologies and consequent scope for increased economic returns, the diversification in favour of such crops will be the future focus. Many other related technologies and their adoption will also inject an added dimension in crop diversification. Decision support systems, governmental policies, geographic information system, application of information technology leading to market information etc., will also lead to crop diversification primarily on economic considerations.

Research and Developmental Support

Future agriculture will be much more knowledge and skill based rather than the traditional subsistence agriculture. In the wake of globalisation and opening up of the global market, there will be much more opportunity for entrepreneurship development in agriculture. This also calls for paradigm shifts in research and technology development and also the transfer of technology for successful crop diversification.

The research system not only needs to address the issues connected with continuance and indulgence and knowledge in the areas of emerging technologies but also create a cadre of scientists through the continuous upgrade of skills and human resource development. The researchers also need to popularise the technologies, impart knowledge and skills to the extension functionaries for the transfer of technologies to the farmers. This knowledge-based farming will call for much more interaction between the researchers, extension workers and farmers. The fruits of the innovative technologies should reach the farmers at the earliest and also spread in the quickest possible time.

Infrastructure Developments

To sustain and operationalise crop diversification, institutional support is required to the two thirds of the country's crop area which is rainfed. Crop diversification in terms of reducing the risk of rainfed farmers is also very vital to a country like India two thirds of the farmers are also resource poor. However, crop diversification in well endowed area is more of an economic consideration. The National Agricultural Research System with its Crop and Commodity based Institutions, Natural Research Management Based Institutions and State Agricultural Universities are jointly addressing the issues connected with the crop diversification. The government of India has also developed a counter support mechanism through the establishments of Crop Directorates for each of the major crops and groups of crops like Oilseeds and Pulses for developmental and technology transfer focus on each of these crops and commodities. These Directorates act as a coordinating agency between the research and development activities on the one hand and between the Federal Government at the central and Regional Governments at the State level for technology transfer and other promotional activities.

Government Policies

Considering the importance of crop diversification in the overall developmental strategy in Indian agriculture, the government of India has taken several initiatives for agricultural development in general and crop diversification in particular. These initiatives are as follows:

i) Launching a Technology Mission for the Integrated Development of Horticulture in the Northeastern Region: The programme will establish effective linkages between research, production, extension, post-harvest management, processing, marketing and exports and bring about a rapid development of agriculture in the region.

ii) Implementing National Agriculture Insurance Scheme: The scheme will cover food crops and oilseeds and annual commercial and horticulture crops. Small and marginal farmers are eligible for 50 percent subsidy under the Scheme.

iii) Operationalising Technology Mission on Cotton: The Technology Mission will have separate Mini-Missions on technology generation, product support and extension, market infrastructure and modernisation of ginning and pressing units.

iv) Provision of Capital Subsidy of 25 percent for construction/modernisation/expansion of cold storages and storages for horticultural produce.

v) Creation of Watershed Development Fund: At the National level for the development of Rainfed lands.

vi) Infrastructure Support for Horticultural Development with emphasis on Post-harvest Management.

vii) Strengthening Agricultural Marketing: Greater attention to be paid for development of a comprehensive, efficient and responsive marketing system for domestic marketing as well as exports by ensuring proper quality control and standardisation.

viii) Seed Crop Insurance: A pilot scheme on Seed Crop Insurance has been launched which will cover the risk factor involved in production of seeds.

ix) Seed Bank Scheme: About 7-8 percent of certified seeds produced in the country will be kept in buffer stock to meet any eventualities arising out of drought, floods or any other form of natural calamities.

x) Cooperative Sector Reforms: Amendment to the National Cooperative Development Corporation (NCDC) Act, 1952, and Replacement of the Multi-State Cooperative Societies (MSCS) Act, 1984.

All these measures will lead to crop diversification and increase the production and productivity of crops.

4

Saline Water in Crop Production

The primary objective of agriculture is to provide the food and fibre needs of human beings. These needs increase as the population increases. The population of the developing countries, which is presently over three-quarters of the worlds total, is projected to increase by about 2.0 percent per year during the last decade of this century and to account for about 90 percent of the expected increase in global population. These growth rates will require an increase in agricultural production of about 40 to 50 percent over the next thirty to forty years, in order to maintain the present level of food intake; a 20 and 60 percent increase for developed and developing countries, respectively.

Growth in crop production can come from increases in arable land, cropping intensity and yield per unit area of cropped land. Nearly two-thirds of the increase in crop production needed in the developing countries in the next decade must come from increases in average yields, a fifth from increases in arable lands, and the balance from increases in cropping intensity. About two-thirds of the increase in arable lands is expected to come from the expansion of irrigation. Thus it is concluded that the needed increases in food production in developing countries must come primarily from existing cropland, mostly irrigated land.

Irrigation has already played a major role in increasing food production over the past fifty years. The world's irrigated land was 8 million hectares in 1800, 48 million hectares in 1900, 94 million hectares in 1950 198 million hectares in 1970, and about 220 million hectares in 1990. About three-quarters of the irrigated land is presently in the developing countries. In these countries, almost 60 percent of the production of major cereals derived from irrigation. Irrigated land presently accounts for 15 percent of the cultivated land but produces 36 percent of the world's food.

Expansion in irrigation needs to be 2.25 percent per year in order to meet food needs by the year 2000. However, the present rate of expansion in irrigation has recently slowed to less than 1 percent per year. The reasons for this slowing down in expansion rate are many. Among them are the high costs of irrigation development and the fact that much of the suitable land and water supplies available for irrigation have already been developed; progressively more expensive and socio-economically less favourable areas are left for further expansion.

Water is the limiting constraint for almost 600 million hectares of potentially suitable arable land. Also, the overall performance of many irrigation projects has been less than expected due to inadequate operation, maintenance and inefficient management. It is not unusual to find that less than 60 percent of the water diverted or pumped for irrigation is actually used in crop transpiration. Furthermore, improper irrigation causes environmental and ecological problems.

Agricultural production systems are limited by the capacity of the associated ecosystems to sustain their natural properties, even though advances in agricultural technology have reduced our dependency somewhat in this regard. The relationship between sustainable agriculture and the environment is one of complimentarity and interdependence. In many locations around the world, strains upon the environment are occurring increasingly and concern is mounting about the sustainability of irrigated agriculture with respect to waterlogging, salinisation, erosion, desertification, loss of biological diversity, water-borne diseases, adverse effects of potentially toxic agricultural chemicals upon human health and the biota of associated eco-systems.

The use of sophisticated agricultural practices has had, so far, a net beneficial effect upon agricultural production, human welfare, nutrition and health. But mismanagement and overuse have the potential to overwhelm the ability of natural processes to "absorb" these practices. A critical challenge facing most countries is to halt and reverse the present extent of environmental degradation resulting from excessive exploitation of natural resources, especially those manifested in desertification, soil erosion, waterlogging, and soil and water salinisation, in order to ensure the needs of future generations.

Presently, 5 to 7 million hectares of arable land are being lost every year through soil degradation. The projected loss by the year 2000 is 10 million hectares annually. By the year 2000, productivity of about one-third of the world's arable land may be severely impaired by excessive erosion. The future expansion of food production will be increasingly dependent upon sound irrigation and water management and upon the concurrent maintenance of the present agricultural resource base and the environment - two of the most challenging tasks facing mankind today.

Expansion of irrigated agriculture could contribute significantly towards achieving and stabilising food and fibre needs. However, new water supplies for such expansion are limited. Irrigated agriculture is already the largest consumer of developed water resources. At the same time, drainage return from irrigated lands is one of the major causes of waterlogging and of water pollution.

Water availability for irrigation could be enhanced through judicious and proper use of saline water and the recycling of drainage waters for irrigation. Considerable amounts of such water are available in various places in the world. Waters generally classified as unsuitable for irrigation can, in fact, be used successfully to grow crops without long-term hasardous consequences to crops or soils, with the use of improved farming and management practices. The development of crops with increased salt tolerance and the adoption of new crop and water management strategies will further enhance and facilitate the use of saline waters for irrigation and crop production, while keeping soil salinity from becoming excessive. The reuse of drainage waters for irrigation will also help to conserve water and to minimise the hasardous effects of irrigation on the environment and ecology.

The development of appropriate practices for the use of saline waters for irrigation requires an adequate understanding of how salts affect waters, soils and plants. But, the sustainability of a viable, permanent irrigated agriculture, especially with the use of saline irrigation waters, requires much more. It requires the implementation of appropriate management practices to control salinity, not only within the irrigated fields, but also within irrigation projects and geohydrologic systems.

It is important to remember that most waterlogging and salinity problems presently existing in major irrigation projects throughout the world have resulted with the use of "good quality" irrigation waters.

Hence, it may be argued that the major causes of salinity problems presently being generally encountered in typical irrigation projects must first be avoided, if more saline than normal waters are to be used successfully for irrigation, since such use may increase the likelihood of salinity problems in a given field. On the other hand, reuse of drainage waters for irrigation can help reduce overall the drainage, waterlogging and salt-loading problems that occur, especially at the project or river basin scales and, hence, can result in a net decrease in the totality of irrigation-induced and salinity-related problems, including environmental pollution. In any case, it is imperative that management practices for the control of soil and water salinity at such scales be considered an essential part of the management requirements for using saline waters for irrigation. This requires the following:

- — that the seriousness of salinity-related environmental problems and the vulnerability of irrigated lands to waterlogging and salination be sufficiently recognised;
- — that the processes contributing to these problems and the effects of salts on soils and plants be understood;
- — that the salinity conditions and trends of the irrigated lands and associated water resources be routinely assessed using appropriate measurement and monitoring techniques that provide meaningful and timely information;
- — that salinity-related problems be properly diagnosed using appropriate criteria and standards;

— that future conditions of soil and water salinity be adequately predicted using appropriate prognostic techniques; and

— that the viability of the irrigated agriculture and associated water resources be sustained by implementing effective long-term control measures.

Saline Water Resources

Because the suitability of a saline water for irrigation is so dependent upon the conditions of use, including crop, climate, soil, irrigation method and management practices, water quality classifications are not advised for assessing water suitability for irrigation. However, for the purpose of identifying the levels of water salinities for which these guidelines are intended, it is useful to give a classification scheme.

Only very tolerant crops can be successfully produced with waters that exceed about 10 dS/m in EC. Few generally-used irrigation waters exceed about 2 dS/m in EC. Many drainage waters, including shallow groundwaters underlying irrigated lands, fall in the range of 2-10 dS/m in EC. Such waters are in ample supply in many developed irrigated lands and have good potential for selected crop production, though they are often not used in this regard and are more typically discharged to better quality surface waters or to waste outlets. It is the use of such saline waters that is the major focus of these guidelines.

Reuse of second-generation drainage waters for irrigation is also sometimes possible and useful, especially for purposes of reducing drainage volume in preparation for ultimate disposal or treatment. Such waters will generally have ECs in the range 10-25 dS/m. Thus, they too are considered in these guidelines, though to a much lesser degree because the "crops" that can be grown with them are atypical and much less experience exists upon which to base management recommendations and to develop guidelines. Very highly saline waters and brine are beyond the scope of these guidelines and their uses for crop production are therefore not discussed herein.

In practical agricultural use, a common source of saline water is groundwater. Salinity of groundwater can be man-induced or natural. In many areas, saline and fresh subsurface waters exist in close proximity. When fresh groundwater is pumped from aquifers that are in hydraulic connection with seawater, the change in gradients as a result of pumping may result in a flow of salt water from the sea towards the well. This is called seawater intrusion.

Upconing is another mechanism by which groundwater could become brackish. Upconing refers to a situation where a well, located close enough to saline water underlying freshwater, is pumped at a rate sufficient to cause the salt water to be drawn into the well in an upward shaped cone or mound. It has been estimated that in the USA over two-thirds of the continental area are underlain by saline groundwater that could intrude on freshwater supplies as a result of upconing.

There are also natural causes of salinity. Numerous investigators have noted that water within sedimentary strata becomes increasingly saline with an increase in depth. In general, the sequence noted is sulphate-rich water near the surface, saline bicarbonate water at an intermediary level and more concentrated chloride water at greater depths. There are several mechanisms by which water trapped in sedimentary rocks can be altered into saline water. One of these is the solution of sediments and rocks.

In coastal regions, surface water sources can become saline due to the tidal influence of the sea. As the high tide moves into the coastal area, seawater moves into streams and drainage canals and travels inland. This upstream migration of seawater alters the quality of water in affected streams and drainage canals significantly. This phenomenon is also observed during times of drought.

Another important source of saline water is drainage effluent from irrigated areas. Drainage water, once thought of as wastewater, is now used in many countries for irrigation. The salinity levels vary, but often the salt levels are higher than those of conventional primary irrigation water sources. Reuse of drainage effluent is important when the supply of good quality irrigation water is limited, and it is also an efficient means of reducing water pollution.

The use of saline drainage water in Egypt was reported by Abu-Zeid. About 2.3 thousand million m3 of drainage wastewater are discharged annually to the Mediterranean Sea via return to the Nile River in Upper Egypt; 12 thousand million m3 are discharged directly into the sea and northern lakes; 2 to 3 thousand million m3 are used for irrigating about 405 000 ha of land. About 75 percent of the drainage water discharged into the sea has a salinity of less than 3000 mg/l. The policy of the Government of Egypt is to use drainage water directly for irrigation if its salinity is less than 700 mg/l; to mix it 1: 1 with Nile water (180 to 250 mg/l) if the concentration is 700 to 1500; or 1: 2 or 1: 3 with Nile water if its concentration is 1500 to 3000 mg/l; and to avoid reuse if the salinity of the drainage water exceeds 3000 mg/l. The potential disadvantages of such blending are discussed later.

Drainage water is used for crop production on many farms in California, USA. For example, saline subsurface drainage water is blended with Delta-Mendota Canal water in the Broadview Water District of California to form blended water of a salinity equivalent to 3.2 dS/m and since 1956 is used to grow a variety of crops. Over time, the cropping pattern in this district has changed as the water quality has decreased. Crops now grown are mostly cotton, barley and alfalfa.

The use of brackish groundwater is reported from Tunisia, India and Israel. De-Malach *et al.* state that in the central Negev of Israel, sugarbeet is grown with saline groundwater of EC = 4.4 dS/m under sprinkler irrigation. Gupta has treated the subject of saline water use in India comprehensively. He reported that the salinity level of the Ganges river in India is very low and average total dissolved salt concentration is less than 200 mg/l. However, there are specific stretches or locations along the river system where salinity level increases due to hydrologic as well as human-induced activities. In the deltaic region of the Ganges river in West Bengal, which comes under tidal influences, the salinity of the water can rise to greater than 10 times the average salinity of the river.

Water Quality Assessment

Salts exert both general and specific effects on plants which directly influence crop yield. Additionally, salts affect certain soil physico-chemical properties which, in turn, may affect the suitability of the soil as a medium for plant growth. The development of appropriate criteria and standards for judging the suitability of a saline water for irrigation and for selecting appropriate salinity control practices requires relevant knowledge of how salts affect soils and plants.

Effects of Salts on Soils

The suitability of soils for cropping depends heavily on the readiness with which they conduct water and air (permeability) and on aggregate properties which control the friability of the seedbed (tilth). Poor permeability and tilth are often major problems in irrigated lands. Contrary to saline soils, sodic soils may have greatly reduced permeability and poorer tilth. This comes about because of certain physico-chemical reactions associated, in large part, with the colloidal fraction of soils which are primarily manifested in the slaking of aggregates and in the swelling and dispersion of clay minerals.

To understand how the poor physical properties of sodic soils are developed, one must look to the binding mechanisms involving the negatively charged colloidal clays and organic matter of the soil and the associated envelope of electrostatically adsorbed cations around the colloids, and to the means by which exchangeable sodium, electrolyte concentration and pH affect this association. The cations in the "envelope" are subject to two opposing processes:

- they are attracted to the negatively-charged clay and organic matter surfaces by electrostatic forces, and
- they tend to diffuse away from these surfaces, where their concentration is higher, into the bulk of the solution, where their concentration is generally lower.

The two opposing processes result in an approximately exponential decrease in cation concentration with distance from the clay surfaces into the bulk solution. Divalent cations, like calcium and magnesium, are attracted by the negatively-charged surfaces with a force twice as great as monovalent cations like sodium. Thus, the cation envelope in the divalent system is more compressed toward the particle surfaces. The envelope is also compressed by an increase in the electrolyte concentration of the bulk solution, since the tendency of the cations to diffuse away from the surfaces is reduced as the concentration gradient is reduced.

The associations of individual clay particles and organic matter micelles with themselves, each other and with other soil particles to form assemblages called aggregates are diminished when the cation "envelope" is expanded and are enhanced when it is compressed. The like-electrostatic charges of the particles which repel one another and the opposite-electrostatic charges which attract one another are relatively long-range in effect. On the other hand, the adhesive forces, called Vanderwaal forces, and chemical bonding reactions involved in the particle-to-particle associations which bind such units into assemblages, are relatively short-range forces. The greater the compression of the cation "envelope" toward the particle surface, the smaller the overlap of the "envelopes" and the repulsion between adjacent particles for a given distance between them. Consequently, the particles can approach one another closely enough to permit the adhesive forces to dominate and assemblages to form.

The phenomenon of repulsion between particles causes more soil solution to be imbibed between them. Because clay particles are plate-like in shape and tend to be arranged in parallel orientation with respect to one another, swelling reduces the size of the inter-aggregate pore spaces in the soil and, hence, permeability. Swelling is primarily important in soils which contain substantial amounts of expanding-layer phyllosilicate clay minerals and which have ESP values in excess of about 15. The reason for this is that, in such minerals, the sodium ions in the pore fluid are first. attracted to the external surfaces of the clay plate. Only after satisfying this do the sodium ions occupy the space between the parallel platelets of the oriented and associated clay particles of the sub-aggregates (called domains) where they create the repulsion forces between adjacent platelets which lead to swelling.

Dispersion and slaking can occur at relatively low ESP values, provided the electrolyte concentration is sufficiently low. The packing of aggregates is more porous than that of individual particles or subaggregates, hence permeability and tilth are better in aggregated conditions. Repulsed clay platelets or slaked subaggregate assembles can lodge in pore interstices, also reducing permeability. Thus, soil solutions composed of high solute concentrations (salinity), or dominated by calcium and magnesium salts, are conducive to good soil physical properties. Conversely, low salt concentrations and relatively high proportions of sodium salts adversely affect permeability and tilth. High pH also adversely affects permeability and tilth because it enhances the negative charge of soil clay and organic matter and, hence, the repulsive forces between them.

During an infiltration event, the soil solution of the topsoil is essentially that of the infiltrating water and the exchangeable sodium percentage is essentially that pre-existent in the soil. Because all water entering the soil must pass through the soil surface, which is most subject to loss of aggregation, topsoil properties largely control the water entry rate of the soil. These observations taken together with knowledge of the effects of the processes discussed above explain why soil permeability and tilth problems must be assessed in terms of both the salinity of the infiltrating water and the exchangeable sodium percentage and the pH of the topsoil.

Decreases in the infiltration rate (IR) of a soil generally occur over the irrigation season because of the gradual deterioration of the soil's structure and the formation of a surface seal created during successive irrigations. IR is even more sensitive to exchangeable sodium, electrolyte concentration and pH than is hydraulic conductivity. This is due to the increased vulnerability of the topsoil to mechanical forces, which enhance clay dispersion, aggregate slaking and the movement of clay in the "loose" near-surface soil, and to the lower electrolyte concentration that generally exists there, especially under conditions of rainfall. Depositional crusts often form in the furrows of irrigated soils where soil particles suspended in water are deposited as the water flow rate slows or the water infiltrates. The hydraulic conductivity of such crusts is often two to three orders of magnitude lower than that of the underlying bulk soil, especially when the electrolyte concentration of the infiltrating water is low and exchangeable sodium is relatively high.

Environmental and Ecological Aspects

The world's natural resource base for food production has already been weakened and the likely additional strain of the expected increase in population and agricultural activity needed to feed it are posing a threat to the prospects of sustainable development in many countries. It is pertinent at this stage to define sustainable agricultural development: "Sustainable agriculture is the management and conservation of the natural resource base, and the orientation of technological and institutional change in such a manner as to ensure the attainment and continued satisfaction of human needs for present and future generations. Such sustainable development in the agriculture, forestry and fisheries sectors conserves land, water, plant and animal genetic resources, is environmentally non-degrading, technically appropriate, economically viable and socially acceptable".

Environmental stress is often the result of the excessive demand for scarce natural resources and the related pollution of the land and water generated by over-development and by poverty. The latter occurs when the poor degrade and destroy their immediate environment as they over-use marginal lands for agriculture and dispose of wastes without treatment to common water supplies in order to meet their living needs. Hence an objective of soil and water conservation must also be to create an economic base which makes it more profitable to conserve and protect resources than to destroy them.

There are a number of potentially undesirable impacts on the environment, as well as on the economic and social components of society, caused by improper irrigation which must be considered if agricultural production is to be sustained, even more so if it is to be expanded by the use of saline waters. These impacts can potentially have far-reaching consequences on present as well as future generations and, hence, can affect the very sustainability of irrigated agriculture.

In a typical irrigation project ,water is diverted from the source and transported through a system of canals to irrigate the cropland. Part of the resulting drainage water is collected and discharged into a nearby stream by means of a system of collector and disposal drains. In this particular project, the irrigation water is low in salinity, crop yields are good and the farmers are profiting. No immediate threat of salinisation or waterlogging is evident within the project itself. However, as a result of project activities:

— the area immediately below the project, which is a nature reserve, has become waterlogged and salinised due to the build-up of a shallow water table there caused by excessive on-farm deep-percolation and seepage of drainage water from the collector and disposal drains within the project;

— the stream into which the drainage from the project is discharged has become polluted with salts and agrochemicals to the point that is no longer suitable for drinking and other domestic purposes by a community in the downstream area;

— the groundwater beneath the project has also become polluted because the subsurface drains do not fully intercept the downward flow of percolated water from the irrigated land. This drainage water is high in salts, nitrates, selenium, boron, pesticides and some other agrochemicals and is a potential health hazard to the people who are using the groundwater for domestic purposes;

— the natural vegetation of the reserve land has undergone undesirable changes in its extent and composition caused by waterlogging and salinisation of the area and, as a consequence, the wildlife population has been diminished and altered in its makeup;

— the water birds which were attracted to the wetland habitat are dying due to selenium toxicity;

— fishermen and hunters who have consumed the fish and game of this wetland and preserve are suffering chronic health problems due to excessive consumption of high selenium;

— the drainage canals and associated wetlands have become breeding sites for mosquitoes; as a result malaria outbreaks are occurring in the project area.

This hypothetical example, albeit an exaggerated one, illustrates the multitude of potential environmental, ecological, health and social problems that can and do sometimes arise as a result of improperly planned and managed irrigation and drainage systems. The use of saline waters for irrigation can either accentuate or help mitigate these problems. Most of the problems depicted in this hypothetical situation can be prevented or greatly minimised with proper design and operation of the irrigation and drainage systems. Implementing an appropriate means of disposing of the saline drainage effluent resulting from irrigation is very important in this regard.

There are at least four major environmentally-related potential hasards associated with irrigation in general and with the use of more saline waters in particular. They are: loss in soil productivity due to salinity and waterlogging, pollution of associated water resources with salts and toxicants by drainage, damage to the associated ecosystems and increased risk to public health resulting from water pollution and waterlogging.

Soil Degradation

Large and increasing proportions of the world's irrigated land are deleteriously affected by waterlogging and excessive salinity. While the exact area affected is not known, it is estimated that approximately 25 percent of the world's irrigated land is damaged by salinisation. Some claim that up to 50 percent of the world's irrigated land may be affected by salt. Certainly no continent is free from salt-affected soils. Serious salt-related problems occur within the boundaries of at least seventy-five countries. Countries with notable salinity problems include Australia, China, Egypt, India, Iraq, Mexico, Pakistan, the republics of the ex-Soviet Union, Syria, Turkey, and USA.

A close relationship exists between the depth and salinity of the shallow groundwaters, the soil hydraulic properties and the extent of salt accumulation in soils, especially in natural, semi-arid regions. The major saline regions of the world are generally found in semi-arid and arid and relatively low-lying, poorly drained lands. This is the result of the mobilisation of large quantities of salts by excessive irrigation and leaching and the subsequent accumulation of the salt in localised areas with restricted drainage. Such areas are often found in lower-lying regions of the landscape where the water table is at or near the soil surface, and where the salts have ascended into the soil due to evaporation-driven processes. Restricted drainage may be due to low permeability of the fine-textured soils or to the presence of a shallow groundwater. Shallow groundwaters are often related to topographic position.

The drainage of waters from the higher-elevation regions of valleys and basins may raise the water table in the lower-lying lands so that it is close to the soil surface. Permeability of the soils is typically lower in these basin positions because of the higher content of alluvial clays generally found in basin soils, which impedes the downward movement of water and results in poor drainage. Many irrigation projects are located in these lower lying alluvial- and basin-position areas because of their favourable slopes and closer proximities to easily accessible water supplies.

While salt-affected soils occur extensively under natural conditions, the salt problems of greatest importance to agriculture arise when previously productive cultivated soil becomes salinised as a result of irrigation (so-called secondary salinisation). Human activities have modified (likely have increased) the extent of salt-affected areas considerably by the redistribution of water (hence salt) through irrigation. The development of large-scale irrigation projects, which involves diversions of rivers, construction of large reservoirs and the irrigation of large landscapes, causes large changes in the natural water and salt balances of entire hydrogeologic systems. The impact of irrigation often extends well beyond that of the immediate irrigated area; even neighbouring nations can be affected.

Water infiltrated into the soil in excess of that used by the agricultural crops passes beyond the rootzone. This water often dissolves salts of geologic origin from the soils and underlying substrata and causes waterlogging in lower areas where it accumulates. When this occurs, soluble salts present in the ground are mobilised and transported to the lower areas where they accumulate and over time salinise the groundwaters and the soils in the areas where the water tables approach ground level.

The problems of waterlogging and secondary salinity prevalent in most irrigated lands have resulted from the excessive use of water for irrigation, from inadequate and inappropriate drainage management, and from the discharge of "spent" drainage water into good-quality water supplies which are used elsewhere for crop production. It is not unusual to find that less than 60 percent of the water diverted for irrigation is used in crop transpiration. It is important to note that these problems have occurred even where low-salinity waters have been used for irrigation. Thus it might be argued that the use of saline waters for irrigation can only increase these problems, since more salt will be added to the soils with such waters and relatively more leaching (hence drainage) is required in this case for salinity control of the rootzone. However, paradoxically, such need not be the case.

It should also be understood that some soil and water salination is inevitable with irrigation. Typical irrigation waters may contain from 0.1 to 4 kg of salts per m3 and are generally applied at annual rates of 1.0 to 1.5m. Thus, from 1 to 60 metric tonnes of salt per hectare may be added to irrigated soils annually. The salt contained in the irrigation water is left in the soil as the pure water passes back to the atmosphere through the processes of evaporation and plant transpiration. Therefore, water in excess of evapotranspiration must be applied with irrigation to achieve leaching and to prevent excess salt accumulation in the rootzone. This water must drain from the rootzone. Seepage from delivery canals occurs in many irrigation projects. These drainage and seepage waters typically percolate through the underlying strata, flow to lower elevation lands or waters and frequently cause problems of waterlogging and salt-loading there. Saline soils are formed in such areas through the processes of evaporation. Ground- and surface-waters receiving these drainage and seepage waters typically are increased in salt concentration. Thus the problems of waterlogging and secondary salinisation are related to inefficient irrigation and/or inadequate drainage.

The primary sources of drainage from an irrigation project are bypass water, canal seepage, deep percolation and surface (tailwater) runoff. Bypass water is often required to maintain hydraulic head and adequate flow through gravity-controlled canal systems. It is usually returned directly to the surface water supply and few pollutants, if any, are picked up in this route. Evaporation losses from canals commonly amount to only a small percentage of the diverted water. However, seepage from unlined canals is often substantial. Such seepage typically contributes significantly to high water tables, increases groundwater salinity and phreatophyte growth, and generally increases the amount of the required drainage (and its salinity) from irrigated areas.

The majority of the soil degradation problems related to irrigated agriculture occurring throughout the world are caused by inefficiencies in the distribution and application of irrigation water, the resulting mobilisation and accumulation of excess water and salts in local regions related to hydrogeologic conditions and the return of saline drainage waters to fresh water supplies. The use of saline waters of the levels advocated herein should not result in excessively saline soils nor cause waterlogging with proper management. In fact, the interception of drainage waters percolating below rootzones and their reuse for irrigation should reduce the soil degradational processes associated with excessive deep percolation, salt mobilisation, waterlogging and secondary salinisation that typically operate in irrigated lands. It should also reduce the water pollution problems associated with drainage discharge to good-quality water supplies.

Water Pollution

The role of irrigated agriculture in soil salinisation has been well recognised for hundreds of years. However, it is of relatively recent recognition that salinisation of water resources from agricultural activities is a major and widespread phenomenon of possibly even greater concern to the sustainability of irrigation than is that of the salinisation of soils, *per se.* Indeed, only in the past few years has it become apparent that trace toxic constituents, such as Se, Mo and As, in agricultural drainage waters may cause pollutional problems that threaten the continuation of irrigation in some projects.

Water infiltrated into the soil in excess of that used by the agricultural crops passes beyond the rootzone. This water, together with that deep percolating from canal seepage, often dissolves additional salts from the soils and underlying substrata. Such mobilised salts, when transported to receiving waters, are a source of pollution, as are the salts applied in the irrigation water which have become concentrated in the drainage water through evapotranspiration. These saline drainage waters pollute good-quality receiving waters when they are allowed to mingle with them. Addition potential sources of pollutants from irrigation are the agrochemicals applied to the soils which may also be, in part, mobilised and discharged in the drainage water.

Ecosystem Disturbances

Few data exist on the degree of degradation of associated ecosystems which can be caused by irrigation, especially with saline waters. This deficiency is due to both the lack of effort that has been made to acquire such information for vast areas of the world and the incomplete understanding of how many of the ecological systems are affected by waterlogging and salinity. The task is made more difficult by the absence of a practical means to monitor changes in large irrigated landscapes systems and associated environments in response to developmental factors.

A real example may serve even better to demonstrate the profound effects irrigation and drainage, especially the effects of saline drainage water disposal, may have upon ecological systems and, in turn, their impact on entire irrigation projects. An example of such a mutual dilemma is the Westside area of the San Joaquin Valley of California and the Kesterson Wildlife Refuge, as summarised by San Joaquin Valley Drainage Program. Before development, the native habitat of the San Joaquin Valley was a lush patchwork of aquatic wetland, riparian forest and valley savannah and it teemed with an abundance and diversity of fish and wildlife found nowhere else in the USA. Grizzly bear, elk, antelope, deer, wolves, quail, geese and a multitude of species of migratory birds, especially waterfowl and shorebirds, populated the Valley. The streams and rivers abounded with trout, salmon and steelhead. Now after about one hundred and fifty years of settlement and the development of irrigated agriculture in the Valley, the quantity and quality of the ecology has been markedly altered. Dams now block most of the major streams to anadromous fish. Impoundments and diversions of the streams for irrigation have depleted the streams of most of their flow, while lack of recharge and discharges of drainage waters to them have increased the salt concentrations of the remaining flow.

The change in habitat has been immense. The Central Valley of California has lost, mostly to agriculture, over 91 percent of its original 4 million plus acres of marsh land. The two major inland lakes (Tulare and Buena Vista) which were once the largest freshwater lakes in the western USA are now farmland. In the San Francisco Bay, which was the outlet for the San Joaquin River and most of the Valley's streams, the water surface has been reduced by 41 percent. Riparian wetlands have been reduced statewide to less than 2% of their original area.

As a consequence of these changes in land use, tremendous losses in native habitat have occurred. Fish and wildlife populations are a fraction of what they were originally. Still substantial populations (about 7-8 million ducks and geese) winter in the Valley. However, where once they found about 105 300 hectares of marsh, they now find only 2025 hectares. Where once they could land on 243 000 hectares of freshwater lakes, they now find only 2835 hectares of saline evaporation ponds.

These drastic reductions in the area of native habitat have resulted in population declines in a number of species and plants endemic to the Valley. Several Valley species have become extinct and others are listed as endangered by the Federal or State Governments. Even though irrigated agriculture has nearly completely altered the original ecology and diversity of the San Joaquin Valley, a new ecological concern has recently emerged to threaten the very existence of continued irrigation in a substantial fraction of the San Joaquin Valley.

Because of the occurrence of waterlogging and a lack of a final outlet for drainage water disposal in much of the San Joaquin Valley, evaporation ponds were created as local outlets for "waste" disposal from irrigation. One such pond (the so-called Kesterson Reservoir) was constructed in 1975 to operate as a storage and flow regulating facility as part of a proposed drainage canal planned to discharge ultimately to the San Francisco Bay and to serve simultaneously as a wildlife refuge. Because of concerns about potential environmental impacts (nitrates and pesticides, primarily) of the disposal of this agricultural drainage on the Bay, construction of the canal ceased in 1978 and the Kesterson Reservoir (486 hectares) became the terminus of the drainage canal serving 3240 hectares of irrigated land and, effectively, an evaporation pond.

At Kesterson, contaminants in the drainage water, specifically selenium at about 35 parts per billion, built up in the food chain, accumulated in the fish and birds using the "pond" and manifested itself by 1982 in gross deformities, reproduction failures and deaths of waterfowl. As a result, in 1985 the Kesterson Reservoir was closed to drainage and the drainage outlets from the source, the Westland Irrigation District, were sealed. Some 2800 hectares of additional evaporation ponds exist in the Valley and another 11 300 hectares are under consideration. However, because of the concerns about the effects of these ponds on the waterfowl, their future is in doubt.

Based on levels of selenium found in a survey of fish and wildlife in the regions of the ponds, health warnings have been issued to avoid or restrict consumption of wild plants, fish and/or wildlife from several areas of the San Joaquin Valley.

Numerous studies and considerable funds have been dedicated to finding a feasible and acceptable solution to the mutual dilemma of finding a means of drainage water disposal from the irrigated lands of the San Joaquin Valley and of sustaining the 320 000 hectares of irrigated land now being threatened by waterlogging and salinity while simultaneously protecting the water quality of the surface and groundwaters, and remaining associated ecological habitats (largely wildlife refuges) of the region.

This example illustrates the new concern about the environment and ecology that is developing worldwide and the new more holistic approach that must be undertaken to balance developmental, environmental and ecological needs. In the case of the San Joaquin Valley "drainage" problem, the approach being undertaken involves a series of programmes. Firstly, source control through the implementation of more efficient irrigation systems and practices are being undertaken to conserve water and reduce deep percolation. Reuse of the unavoidable drainage waters through a succession of crops of increasing salt tolerance, including eucalyptus and halophyte species, is also being implemented so as to reduce drainage water volumes and conserve water, while producing useful biomass. Conjunctive use of saline groundwater and surface water is being considered to aid in lowering water tables, hence reducing drainage disposal need, and conserving water. Treatment of drainage water and various means of ultimate disposal of the unusable final drainage effluent through deep aquifer injection and ecologically safe evaporation ponds and its release during high stream-flow periods are also under consideration. Lastly, release of freshwater supplies to refuge areas and the retirement of irrigated land deemed the major source of the pollutional problems are also being considered. All of these so-called "in-valley" solutions are being put ahead of the construction of a master drain and ocean disposal in keeping with the philosophy of dealing with the problem at the source and in making the "polluters" pay the costs of pollution that they cause rather than allowing them to discharge their wastes at the expense of others.

Management Practices for Safe Use of Saline Water

While irrigated agriculture has greatly increased crop productivity, inappropriate and inefficient irrigation has wasted water, polluted surface water and groundwater, damaged productivity and altered the ecology of vast areas of land. Contamination of water supplies by irrigation is, in many places, posing health risks and drastically increasing the costs of treating waters for domestic and industrial uses. Surface and groundwaters in many areas are being contaminated by salts, fertilisers, herbicides and pesticides. Toxic chemicals are rendering many developed water supplies unfit for drinking and even for irrigation in some cases. These pollutants also degrade the recreational use and esthetic value of surface waters. At the same time, costly limitations are being placed upon irrigation to reduce its pollutional discharges or to treat its wastes before discharge. Finding a suitable, acceptable place for such discharge is increasingly becoming a major problem in some situations, especially in the developed countries. Blending saline and fresh waters reduces the potential usability of the total water supply. Use of polluted waters for irrigation limits crop production potential, as well as posing some potential health hazards to the consumers of the food.

To overcome the above-described problems, new techniques need to be developed and implemented to reduce excessive water uses and to conserve limited water supplies and better ways must be found to implement existing methods more effectively. Efficiency of irrigation must be increased by the adoption of appropriate management strategies, systems and practices and through education and training. Reuse of wastewater, including the use of drainage water and shallow saline groundwater for crop production, must be made an integral component of irrigation water management, water conservation and environmental protection programmes.

Effective salinity control measures must be implemented to sustain irrigated agriculture and to prevent pollution of associated water resources. Such measures must be chosen with recognition of the natural processes operative in irrigated, geohydrologic systems, not just those on-farm, and with an understanding of how they affect the quality of soil and water resources, not just crop production. Some practices can be used to control salinity within the crop rootzone, while other practices can be used to control salinity within larger units of management, such as irrigation projects, river basins, etc. Additional practices can be used to protect off-site environments and ecological systems - including the associated surface and groundwater resources. The "on-farm" practices usually consist of agronomic and engineering techniques applied by the farmer on a field-by-field basis. The "district-wide" or "larger organisational basis" practices generally consist primarily of engineering structures for water control and systems for the collection, reuse, treatment and/or disposal of drainage waters.

There is usually no single way to achieve safe use of saline water in irrigation. Many different approaches and practices can be combined into satisfactory saline water irrigation systems; the appropriate combination depends upon economic, climatic, social, as well as edaphic and hydrogeologic situations. Thus, no procedures are given here for selecting "the" appropriate set of practices for different situations. Rather, some important goals, principles and strategies of water, soil and crop management practices that should be considered in the use of saline water for irrigation are presented as guidelines.

Management Principles

Salinity management constitutes an important aspect of safe use of saline water irrigation. This requires an understanding of how salts affect plants and soils, of how hydrogeologic processes affect salt accumulation, and also of how cropping and irrigation activities affect soil and water salinity.

To prevent the excessive accumulation of salt in the rootzone from irrigation, extra water (or rainfall) must, over the long term, be applied in excess of that needed for ET and must pass through the rootzone in a minimum net amount. This amount, in fractional terms, is referred to as the "leaching requirement"(Lr, the fraction of infiltrated water that must pass through the rootzone to keep salinity within acceptable levels). In fields irrigated to steady-state conditions with conventional irrigation management, the salt concentration of the soil water is essentially uniform near the soil surface regardless of the leaching fraction (LF, the fraction of infiltrated water that actually passes through the root-zone) but increases with depth as LF decreases. Likewise, average rootzone salinity increases as LF decreases; crop yield is decreased when tolerable levels of salinity are exceeded. Methods to calculate the leaching requirement and to predict crop yield losses due to salinity effects were described previously. Once the soil solution has reached the maximum salinity level compatible with the cropping system, at least as much salt as is brought in with additional irrigations must be removed from the rootzone; a process called "maintaining salt balance."

To prevent waterlogging and secondary salination, drainage must remove the precipitation and irrigation water infiltrated into the soil that is in excess of crop demand and any other excessive water (surface or subsurface) that flows into the area; it must provide an outlet for the removal of salts that accumulate in the rootzone in order to avoid excessive soil salinisation, and it must keep the water table sufficiently deep to permit adequate root development, to prevent the net flow of salt-laden groundwater up into the rootzone by capillary forces and to permit the movement and operation of farm implements in the fields.

Artificial drainage systems may be used in the absence of adequate natural drainage. They are essentially engineering structures that control the water table at a safe level according to the principles of soil physics and hydraulics. The water table depth required to prevent a net upward flow of water and salt into the rootzone is dependent on irrigation management and is not single-valued as is commonly assumed. Methods to calculate drainage requirements are given elsewhere.

The time-averaged level of rootzone salinity is affected by the degree to which the soil water is depleted between irrigations, as well as by the leaching fraction. As the time between irrigations is increased, soil water content decreases as the soil dries, and the matric and osmotic potentials of the soil water decrease as salts concentrate in the reduced volume of water. Water uptake and crop yield are closely related to the time and depth averaged total soil water potential, i.e. matric plus osmotic.

As water is removed from a soil with non-uniform salinity distribution, the total water potential of the water being absorbed by the plant tends to approach uniformity in all depths of the rootzone. Following irrigation, plant roots preferentially absorb water from rootzone depths with high water potential. Normally this means that most of the water uptake is initially from the upper, less saline soil depths until sufficient water is removed to increase the total water stress to a level equal to that in the lower depths. After that water is removed from the deeper, more saline soil depths and the effect of salinity, *per se*, on crop growth is magnified. This implies that:

— forms of irrigation that minimise matric stress, such as drip irrigation, can be used to minimise the harmful effects of irrigating with saline water;

— high leaching fractions can be used to minimise the buildup (hence harmful effects) of high levels of salinity in deeper regions of the rootzone.

The distribution within and the degree to which a soil profile becomes salinised are also functions of the manner of water application, as well as the leaching fraction. More salt is generally removed per unit of leachate with sprinkler irrigation than with flood irrigation. Thus, the salinity of water applied by sprinkler irrigation can be somewhat higher, all else being equal, than that applied by flood or furrow irrigation with a comparable degree of cropping success, provided foliar burn is avoided. The high salt-removal efficiency of sprinkler irrigation may be explained as follows. Solute transport is governed by the combined processes of convection and diffusion; convection is usually the predominant process in flood-irrigated soils.

Differential velocities of water flow can occur within the soil matrix because the pore size distribution is typically non-uniform. This phenomenon is called dispersion. It can be appreciable when flow velocity is high and pore size distribution is large; diffusion often limits salt removal under such conditions. Soils with large cracks and well-developed structure are especially variable in their water and solute transport properties because the large "pores" are preferred pathways for water flow, as are earthworm channels, old root holes, interpedal voids, etc.; most of the flow in flooded soils occurs via these "pores".

Much of the water and salt in the small and intra-aggregate pores is "bypassed" in flood irrigated soils. Flow velocity and water content are typically lower in soils irrigated with sprinklers; hence, bypass is reduced and efficiency of salt leaching is increased. Other soil-related processes also affect salt concentration and transport during the irrigation and leaching of soils. In most arid land soils, the clay particles are dominated by negative charges, which can retard cation transport through adsorption and/or exchange processes. Simultaneously, anions are largely excluded from that part of the pore solution adjacent to the negatively-charged clay surface; this accelerates their relative transport. The borate anion also undergoes adsorption reactions that retard its movement.

The distribution of salts in the soil is also influenced by seedbed shape. Salts tend to accumulate to excess levels in certain regions of the seedbed under furrow irrigation.

Information from this early study shows that seedbed and furrow shape can be designed to minimise this problem. Seed placement and surface irrigation strategies that can also be used to optimise plant establishment under saline conditions are described by Kruse *et al.*. Sprinkler irrigation can be effective in leaching excessive salinity from the top-soil and in producing a favourable low-salinity environment in the upper soil layer which is necessary for the establishment of salt-sensitive seedings. However, other problems are associated with sprinkling of saline water. Saline, "bed-peaks" can be detopped to prevent exposure to emerging shoots. Under drip irrigation, the salt content is usually lowest in the soil immediately below and adjacent to the emitters and highest in the periphery of the wetted zone. Removal of salt that has accumulated in this wetting zone "front" must be addressed in the long-term.

Susceptible crops should not be irrigated with saline water by sprinkler irrigation since their foliage absorbs salts upon wetting. Salts can accumulate in leaves by foliar absorption of such crops until lethal concentrations have been reached. Crop sensitivity to saline sprinkling water is related more to the rate of foliar salt accumulation than to crop tolerance to soil salinity, *per se.* Hence, applications should be made during the night and in a manner to achieve frequent wetting of the leaves in order to minimise foliar absorption of salts when irrigating with saline waters by sprinkler methods.

The prevalent models of solute reactions and transport in irrigated soils suffer the deficiency of not appropriately representing the effects of the above-described processes that often occur under field conditions. Neither do they adequately account for the distribution uniformity effects of the irrigation application system itself, or of the infiltration uniformity effects resulting from variable soil permeability across the field. Only recently has this problem been approached directly by measuring, on a large scale, solute distributions in field soil profiles. The results to date indicate that as yet no suitable method to quantify and integrate the effects of these processes on a field basis exists. It is probable that alternative modelling approaches, like that proposed by Corwin and Waggoner, may help in this regard.

Some soil and water salination is inevitable with irrigation; the salt contained in the irrigation water remains in the soil as the pure water passes back to the atmosphere through the processes of evaporation and plant transpiration. Therefore, water in excess of evapotranspiration must be applied with irrigation to achieve leaching and prevent excess salt accumulation. This water must drain from the rootzone. Seepage from delivery canals also occurs in many irrigation projects. These drainage and seepage waters percolate through the underlying strata, flow to lower elevation lands or waters and frequently cause problems there of waterlogging and salt-loading. Saline soils typically are formed in such lands through the processes of evaporation. Ground and surface waters receiving these drainage and seepage waters typically are increased in salt concentration.

The primary sources of return flow from an irrigation project are bypass water, canal seepage, deep percolation, and surface runoff. Bypass water is often required to maintain hydraulic head and adequate flow through a gravity-controlled canal system. It is usually returned directly to the river, and few pollutants, if any, are picked up in this route. Evaporation losses from canals commonly amount to only a small percentage of the diverted water. Seepage from unlined canals is often substantial. It may contribute to high water tables, increase groundwater salinity and phreatophyte growth, and generally increases the amount and salinity of the required drainage from irrigated areas.

Law *et al.* estimated that 20 percent of the total water diverted for irrigation in the USA is lost by seepage from conveyance and irrigation canals. If the water passes through salt-laden substrata or displaces saline groundwater, the salt pickup from this source can be substantial. Canal lining can reduce such salt loading. Closed conduit conveyance systems can minimise both seepage and evaporation losses and ET by phreatophytes. The closed conduit system also provides the potential to increase project irrigation efficiency and to thus lower salt loading. Reducing the volume of water applied for irrigation proportionately reduces the amount of salt added and the amount needed to be removed by leaching. Minimising the leaching fraction maximises the precipitation of applied Ca, HCO_3, and SO_4 salts as carbonates and gypsum minerals in the soil, and it minimises the "pickup" of weathered and dissolved salts from the soil. The salt load from the rootzone can be reduced from about 2 to 12 tons/ha per year by reducing LF from 0.3 to 0.1 .

Minimising leaching may or may not reduce salinity degradation of the receiving water where the drainage water is not intercepted and isolated and is returned to the associated surface or groundwater. A reduction of degradation will generally occur where saline groundwaters with concentrations in excess of those of the recharging rootzone drainage waters are displaced into the surface water or where additional salts, other than those derived from the irrigation water, are encountered in the drainage flow path and brought into solution by weathering and dissolution processes.

Groundwaters receiving irrigation drainage water may not always benefit from reduced leaching. With no sources of recharge other than drainage return flow, the groundwater eventually must come to the composition of the drainage water, which will be more saline with low leaching. Reduced leaching slows the arrival time of the leachate. Thus, the groundwater salinity may be lower with reduced leaching for an interim period of time.

For groundwater under-saturated with $CaCO_3$ (unlikely in arid lands) being pumped for irrigation with no recharge other than by drainage return, groundwater will be slightly less saline under low leaching; groundwater saturated with $CaCO_3$ will show no benefit under low leaching; and groundwater saturated with $CaCO_3$ and nearing saturation with gypsum will show substantial benefit from low leaching. Low leaching management can continuously reduce degradation of the groundwater, only if other sources of high-quality recharge into the basin exist and if flow out of the basin is high relative to drainage inflow.

The extent to which leaching can be minimised is limited by the salt tolerances of the crops being grown, the irrigation system distribution uniformities and the variability in soil infiltration rates. In most irrigation projects, the currently used leaching fractions can be reduced appreciably without harming crops or soils, especially with improvements in irrigation management.

Management for Crop Production

Management practices for the safe use of saline water for irrigation primarily consist of:

— selection of crops or crop varieties that will produce satisfactory yields under the existing or predicted conditions of salinity or sodicity;

— special planting procedures that minimise or compensate for salt accumulation in the vicinity of the seed;

— irrigation to maintain a relatively high level of soil moisture and to achieve periodic leaching of the soil;

— use of land preparation to increase the uniformity of water distribution and infiltration, leaching and removal of salinity;

— special treatments to maintain soil permeability and tilth. The crop grown, the quality of water used for irrigation, the rainfall pattern and climate, and the soil properties determine to a large degree the kind and extent of management practices needed.

Growing Suitably Tolerant Crops

Where salinity cannot be kept within acceptable limits by leaching, crops should be selected that can produce satisfactory yields under the resulting saline conditions. In selecting crops for saline soils, particular attention should be given to the salt tolerance of the crop during seedling development, because poor yields frequently result from failure to obtain a satisfactory stand. Some crops that are salt tolerant during later stages of growth are quite sensitive to salinity during early growth.

Managing Seedbeds

Failure to obtain a satisfactory stand of furrow-irrigated row crops on moderately saline soils is a serious problem in many places. This is because the rate of germination is reduced by excessive salinity. The failures are usually due to the accumulation of soluble salt in raised beds that are "wet-up" by furrow irrigation. Modifications in irrigation practice and bed shape should be used to reduce salt accumulation near the seed. The tendency of salts to accumulate near the seed during irrigation is greatest in single-row, round-topped beds.

Sufficient salt to prevent germination may concentrate in the seed zone even if the average salt content of the soil is moderately low. Thus, such beds should be avoided when irrigating with saline waters using furrow methods, though "decapping" techniques may be used to advantage in this regard. With double-row, flat-topped beds, since most of the salt moves into the centre of the bed, the shoulders are left relatively free of salt, thus seedling establishment may be enhanced by planting on the shoulders of such beds.

Sloping beds are best for saline conditions because the seed can be safely planted on the slope below the zone of high salt accumulation. Such beds should be used, if possible, when furrow irrigating with saline waters. Planting in furrows or basins is satisfactory from the standpoint of salinity control but is often unfavourable for the emergence of many row crops because of problems related to crusting and poor aeration. This method is recommended only for the use of very saline irrigation waters and vigorous, hardy emerging plants. Pre-emergence irrigation by use of sprinklers or special furrows placed close to the seed may be used to keep the soluble salt concentration low in the seedbed during germination and seedling establishment. After the seedlings are established, the special furrows may then be abandoned and new furrows made between the rows, and sprinkling replaced by furrow irrigation.

Careful grading of land makes possible a more uniform application of water and, hence, better salinity control when irrigating with saline water. Barren or poor areas in otherwise productive fields are often either high spots that do not receive enough water for good crop growth or for leaching purposes or low spots that remain too wet for seedling establishment. Lands that have been irrigated one or two years after initial grading usually need to be regraded to remove the surface unevenness caused by the settling of fill material.

Annual crops should be grown after the first grading so that regrading can be performed before a perennial crop is planted. A prior detailed topographic survey could be very helpful to avoid ruining soil properties and in particular removing the surface soil which may be relatively more fertile. Land levelling causes a significant soil compaction due to the weight of the heavy equipment and it is advisable to follow this operation with subsoiling, chiselling and ploughing to break up the compaction and restore or improve water infiltration.

Managing soils under Saline Water Irrigation

Several physical, chemical and biological soil management measures help facilitate the safe use of saline water in crop production. Some important ones in this regard are: tillage, deep ploughing, sanding, use of chemical amendments and soil conditioners, organic and green manuring and mulching.

Tillage is a mechanical operation that is usually carried out for seedbed preparation, soil permeability improvement, to break up surface crusts and to improve water infiltration. If tillage is improperly executed, it might form a plough layer or bring a salty layer closer to the surface. Sodic soils are especially subject to puddling and crusting; they should be tilled carefully and wet soil conditions avoided. Heavy machinery traffic should also be avoided. More frequent irrigation, especially during the germination and seedling stages, tends to soften surface crusts on sodic soils and encourages better stands.

Deep ploughing refers to depths of ploughing from about 40 to 150 cm. It is most beneficial on stratified soils having impermeable layers lying between permeable layers. In sodic soils, deep ploughing should be carried out after removing and reclaiming the sodicity, otherwise it will cause complete disturbances and collapse of the soil structure. Deep ploughing to 60 cm loosens the aggregates, improves the physical condition of these layers, increases soil-water storage capacity and helps control salt accumulation when using saline water for irrigation.

Crop yields can be markedly improved by ploughing to this depth every three or four years. The selection of the right plough types (shape and spacings between shanks), sequence, ploughing depth and moisture content at the time of ploughing should provide good soil tilth and improve soil structure. Special equipment can even invert whole soil profiles or break up substrata as deep as 2.5m that impede deep percolation, so that many adverse physical soil conditions associated with land irrigated with saline water can be modified in order to improve leachability and drainability.

Sanding is used in some cases to make a fine textured surface soil more permeable by mixing sand into it, thus a relatively permanent change in surface soil texture is obtained. When properly done, sanding results in improved root penetration and better air and water permeability which facilitates leaching by saline sodic water and when surface infiltration limits water penetration. The method can be combined with initial deep ploughing.

Chemical amendments are used to neutralise soil reaction, to react with calcium carbonate and to replace exchangeable sodium by calcium. This decreases the ESP and should be followed by leaching for removal of salts derived from the reaction of the amendments with sodic soils. They also decrease the SAR of irrigation water if added in the irrigation system. Gypsum is by far the most common amendment for sodic soil reclamation, particularly when using saline water with a high SAR value for irrigation. Calcium chloride is highly soluble and would be a satisfactory amendment especially when added to irrigation water. Lime is not an effective amendment for improving sodic conditions when used alone but when combined with a large amount of organic manure it has a beneficial effect. Sulphur too can be effective; it is inert until it is oxidised to sulphuric acid by soil micro-organisms. Other sulphur-containing amendments are similarly effective because of the sulphuric acid originally present or formed upon microbial oxidation or hydrolysis.

The choice of an amendment for a particular situation will depend upon its relative effectiveness judged from its improvement of soil properties and crop growth, the availability of the amendments, relative cost involved, handling and application difficulties and time allowed and required for the amendment to react in soil and effectively replace adsorbed sodium.

Attempts have been made to coagulate soil particles and provide deep aeration and better permeability and water infiltration by chemical treatment. Treating the soil with dilute bituminous emulsions can result in effective aggregation, improved aggregate stability and reduced surface crust formation. Water percolation rate is faster in bitumen-treated soil.

Sulphate lignin conditioner can also be used to improve soil structure, and to improve soil permeability. Soil conditioners can have practical applications in seedling establishment when soil is irrigated with saline water of high SAR. Stability of soil aggregates prevents dispersion and formation of deposit crusts and infiltration can be maintained by application of small quantities of organic polyelectrolytes to the soil surface. They can be effective when introduced in the irrigation water or when sprayed over the soil surface.

Mineral fertilisers: Salt accumulation affects nutrient content and availability for plants in one or more of the following ways: by changing the form in which the nutrients are present in the soil; by enhancing loss of nutrients from the soil through heavy leaching or, as in nitrogen, through denitrification, or by precipitation in soil; through the effects of non-nutrient ions on nutrient uptake; and by adverse interactions between the salt present in saline water and fertilisers, decreasing fertiliser use efficiency.

Crop response to fertiliser under saline or sodic conditions is complex since it is influenced by many soil, crop and environmental factors. The benefits expected from using soil management measures to facilitate the safe use of saline water for irrigation will not be realised unless adequate, but not excessive, plant nutrients are applied as fertilisers. The level of salinity may itself be altered by excess fertiliser application as mineral fertilisers are for the most part soluble salts. The type of fertiliser applied, when using saline water for irrigation, should preferably be acid and contain Ca rather than Na taking into consideration the complementary anions present. Timing and placement of mineral fertilisers are important and unless properly applied they may contribute to or cause a salinity problem.

Organic and green manures and mulching: Incorporating organic matter into the soil has two principal beneficial effects of soils irrigated with saline water with high SAR and on saline sodic soils: improvement of soil permeability and release of carbon dioxide and certain organic acids during decomposition. This will help in lowering soil pH, releasing calcium by solubilisation of $CaCO_3$ and other minerals, thereby increasing ECe and replacement of exchangeable Na by Ca and Mg which lowers the ESP. Growing legumes and using green manure will improve soil structure. Green manure has a similar effect to organic manure. Salinisation during fallowing may be severe where a shallow water table exists, since evaporation rates of about 8, 3 and 1 mm/day could occur from the dry surface of fine sandy loam when the water table is kept at 90, 120 and 180 cm from the soil surface, respectively.

Mulching to reduce evaporation losses will also decrease the opportunity for soil salinisation. When using saline water where the concentration of soluble salts in the soil is expected to be high in the surface, mulching can considerably help leach salts, reduce ESP and thus facilitate the production of tolerant crops. Thus, whenever feasible, mulching to reduce the upward flux of soluble salts should be encouraged.

Operating Delivery Systems

Water delivery and distribution systems must be operated efficiently to facilitate the timely supply of water in the right quantities and to avoid waterlogging and salinity build-up in irrigated lands, especially when saline waters are involved. The amount of water applied should be sufficient to supply the crop and satisfy the leaching requirement but not enough to overload the drainage system.

Over-irrigation contributes to the high water table, increases the drainage requirement and is a major cause of salinity build-up in many irrigation projects of the world. Therefore, a proper relation between irrigation, 'eaching, and drainage must be maintained in order to prevent irrigated lands from becoming excessively waterlogged and salt-affected.

Often irrigation water delivery and distribution systems are over-designed, in the absence of reliable data or appropriate methods to predict project water requirements. It is all the more important, when using saline waters, that excessive amounts are not diverted into irrigation schemes as this is likely to cause more damage than excessive amounts of "good quality" water. FAO has developed methods to determine project water requirements based on actual crop water needs, leaching requirements and irrigation efficiencies.

A computer program, called CROPWAT has been developed to calculate crop water requirements and irrigation requirements from climatic and crop data. Further, the program allows the development of irrigation schedules for different management conditions and the calculation of scheme water supply for varying cropping patterns. The program runs on any standard personal computer with a minimum of 360 Kb of memory. The program can be obtained from FAO on request. A complementary computerised database program called CLIMWAT is available to obtain the required climatic data for CROPWAT. CLIMWAT has data from a total of 3262 meteorological stations from 144 countries.

Excessive loss of irrigation water from canals constructed in permeable soil is a major cause of high water tables and secondary salination in many irrigation projects. Such seepage losses should be reduced by lining the canals with impermeable materials or by compacting the soil to achieve a very low permeability. Because the amount of water passing critical points in the irrigation delivery system must be known in order to provide water control and to achieve high water-use efficiency, provisions for effective flow measurement should be made. Unfortunately, many current irrigation systems do not use flow measuring devices and, thus, the farmers operate with limited control and knowledge of the amount of water actually diverted to the farms. In addition, many delivery systems encourage over-irrigation because water is supplied for fixed periods, or in fixed amounts, irrespective of seasonal variations in on-farm needs. Salinity and water table problems are often the result. The distribution system should be designed and operated so as to provide water on demand and in metered amounts as needed to achieve high efficiency and to facilitate salinity control and the use of saline waters for irrigation.

Efficient Irrigation

Improvements in salinity control generally come hand-in-hand with improvements in irrigation efficiency. The key to the effective use of saline irrigation waters and salinity control is to provide the proper amount of water to the plant at the proper time. The ideal irrigation scheme should provide water as nearly continuously as possible, though not in excess, as needed to keep the soil water content in the rootzone within optimum safe limits. However, carefully programmed periods of stress may be needed to obtain maximum economic yield with some crops; cultural practices also may demand occasional periods of dry soil.

Thus, the timing and amount of water applied to the rootzone should be carefully controlled to obtain good water use efficiency and good crop yield, especially when irrigating with saline water. As mentioned above, this requires water delivery to the field on demand which, in turn, requires the establishment of close coordination between the farmer and the entity that distributes the water; it calls for the use of feedback devices to measure the water and salt contents and potentials in the soil and devices to measure water flow in the conveyance systems.

The method and frequency of irrigation and the amount of irrigation water applied may be managed to control salinity. The main ways to apply water are basin flooding, furrow irrigation, sprinkling, subirrigation, and drip irrigation. Flood irrigation is good for salinity control when using saline waters if the land is level, though aeration and crusting problems may occur. Aeration and crusting problems are minimised by using furrow irrigation, but salts tend to accumulate in the beds. If excess salt does accumulate, a rotation of crops and periodic irrigation by sprinkler or flooding should be used as salinity-control measures.

Alternatively, cultivation and irrigation depths should be modified, once the seedlings are well established, to "shallow" the furrows so that the beds will be leached by later irrigations. Irrigation by sprinkling may give better control of the amount and distribution of water; however, the tendency is to apply too little water by this method, and leaching of salts beyond the rootzone may sometimes be accomplished only with special effort. Salinity can be kept low in the seedbed during germination with sprinkler-irrigation, but crusting may be a problem. Emergence problems associated with such crusting may be overcome with frequent light irrigations during this time or by use of special tillage techniques.

Subirrigation with saline water is not generally advisable unless the soil is periodically leached of the accumulated salts by rainfall or by surface applications of low-salinity water. Drip irrigation, if properly designed, is recommended for use of saline irrigation water because it minimises salinity and matric stresses in the rootzone, though salts accumulate in the periphery of the wetted area. As noted earlier, higher levels of salinity in the irrigation water can be tolerated with drip as compared with other methods of irrigation.

Because soluble salts reduce the availability of water in almost direct proportion to their total concentration in the soil solution, irrigation frequency should be increased so that the moisture content and salinity of irrigated soils are maintained as high and low, respectively, as is practicable, especially during seedling establishment and the early stage of vegetative growth, if it can be done without resulting in excessive leaching or insufficient depth of rooting. The most practical way to accomplish this is through use of drip irrigation.

Additional water must be applied, at least occasionally, to leach out the salt that has accumulated during previous irrigations. This leaching requirement depends on the salt content of the irrigation water and on the maximum salt concentration permissible in the soil solution which depends in turn on the salt tolerance of the crop and the manner of irrigation.

Fortunately, much of the needed leaching can be achieved between crops or during pre-irrigation and early growth-stage irrigations when soil permeability is generally relatively high, especially when using low-salinity waters in the cyclic use strategy. The first irrigations provided for the renewal of cropping following a fallow or uncropped period often unavoidably result in relatively high leaching.

Many irrigation practices, especially with flood irrigation, inadvertently result in excess leaching, especially during pre-plant or early-season irrigations before the soil aggregation has slaked and surface soil permeability has diminished. Effects of non-uniform crop stand and cover, soil infiltration rates (permeabilities) and water application and distribution result in generally non-uniform leaching across an irrigated field.

Automated solid-set and centre-pivot sprinklers systems are conducive to good control and uniform distribution of applied water; in principle, trickle irrigation 5s even better. But gravity systems, if designed and operated properly, can also achieve good uniformity. Precision land grading and use of smaller water applications should be used to facilitate the achievement of high uniformity of areal water distribution over the field and infiltration, respectively.

Closed conduits, rather than open waterways should be used for water distribution laterals if possible; they have the advantage of more effective off-on control, in addition to capturing gravitational energy for use in pressurising delivery systems or controls which offer better potential for achieving high irrigation efficiencies.

The most advanced centre pivot irrigation system now used by some farmers in the USA is called the LEPA system - Low Energy Precision Application. In this system the sprinkler and spray nozzles used on the centre pivot systems are replaced with drop-down hoses and low pressure emitters that operate at only 0.3 kg/cm2 and are placed as close as 20-45 cm above the ground. Experience has shown that these systems can reduce evaporation and wind-drift losses to less than 5 percent of the emitted water.

The new LEPA technology became commercially available in 1986 and since then has been adopted by an estimated 1000 farmers. The advantage of LEPA with regard to saline water irrigation is that it can irrigate crops with the right amount of water, avoiding excess and runoff, and minimise foliar damage which is common with saline water irrigation. However, the technique is new, costly and needs to be further developed to reduce costs and make the system simpler for adoption by a wider group of farmers.

In furrow-irrigated areas, furrow length should be reduced in order to improve intake distribution and to reduce tail water runoff. Worstell's multi-set system is useful for such purposes. Surge irrigation techniques can sometimes be used to improve irrigation uniformity in graded furrows.

For tree crops, a low-head bubbler system can be used to provide excellent control and to minimise the pressure requirements and expensive filtration systems. Drip systems, of course, are increasingly being used for permanent crops and high-value annual crops and are well suited for use with saline irrigation waters. All opportunities to modify existing irrigation systems to increase their effectiveness of water and salinity control should be sought and implemented.

A frequent constraint in improving on-farm water use is the lack of knowledge of just when an irrigation is needed and of how much capacity for storage is available in the rootzone. Ways to detect the onset of plant stress and to determine the amount of depleted soil water are prerequisites to supplying water on demand and in the amount needed. Prevalent methods of scheduling irrigation usually do not, but should, incorporate salinity effects on soil-water availability. When irrigating with saline waters, the osmotic component of the soil water potential of the rootzone must be considered in scheduling decisions.

Ideally, irrigation management should have the available soil water near the upper limit during germination and emergence but depleted by about 50 percent, or more, at harvest and should maintain available water within the major rootzone during the early vegetative, flowering and yield formation growth stages at a level which produces no deleterious plant water stress through successive, properly-timed irrigations.

Under saline conditions, some "extra" water must be given for leaching - a minimum commensurate with salt tolerance of the crop being grown, if rainfall is inadequate in this regard, as discussed previously. Some method of assessing the water availability to the crop with sufficient lead time to provide for a water application before significant stress occurs should be used for irrigation scheduling purposes. In addition, the amounts of water needed for replenishment of the depleted soil moisture from the rootzone and for leaching must be determined.

Prevalent methods used to determine the onset of stress include both direct and indirect measurements. Leaf water potential can be measured with a pressure "bomb" and used to determine the onset of stress; however, the method does not give information with which to predict when the stress will occur much in advance of its occurrence, nor does it provide a measure of the amount of water to apply.

Infrared thermometry can also be used to measure plant water stress indirectly which results in the partial closure of leaf stomates and in reduced transpiration rates, causing leaf temperature to rise abnormally above ambient air temperature. This temperature difference can be interpreted in terms of a crop water stress index with which irrigation-need can be assessed. It suffers the same limitation as the leaf water potential method.

Various scheduling methods can be used which are based on sensing depletion of soil-water per se or soil water potential (matric, osmotic or total), or some associated soil or water property, and knowledge of the critical level (the set-point value). Such levels can be ascertained from salt tolerance data by converting threshold ECe values to osmotic potentials and assuming equivalent crop yield loss (also ET loss) would result from total water potential (i.e. assuming the effects of matric and osmotic potentials are equivalent and additive). Matric potential should be measured by any suitable means. Osmotic potential should be determined by one of the methods of salinity measurement described in Rhoades.

Daily potential evapotranspiration can be calculated from measurements of air temperature, humidity, solar radiation and wind or of pan evaporation. The actual evapotranspiration (ETa) can then be estimated from empirically determined, crop coefficients as described by Doorenbos and Kassam. The summation of these daily ETa values can then be used to estimate accumulative soil water depletion and total water potential.

A plot of depletion or water potential versus time is then used to project the need for irrigation. This basic approach can be used based regardless of whether direct measurements of soil water content, or a related parameter, using neutron meters, resistance blocks, time-domain reflectometric (TDR) sensors, four-electrode sensors, or various soil matric potential sensors, etc., are used or estimated from ET methods. All of the methods suffer the limitation of needing to know the critical set-point value for irrigation, which varies with crop type, rooting characteristics, stage of plant growth, soil properties and climatic stress, etc. An estimate of this value can be obtained as described above or by the method of Doorenbos and Kassam.

For saline water, irrigations should be scheduled before the total soil water potential (matric plus osmotic) drops below the level which permits the crop to extract sufficient water to sustain its physiologic processes without loss in yield. Since, typically, the crop's root system normally extracts progressively less water with increasing soil depth, the frequency of irrigations should be determined by the level of total soil water potential in the upper half of the rootzone where the rate of water depletion is greatest. Besides the extent of soil water depletion by ET, determination of the amount of water to apply should also be based on stage of plant development, the salt tolerance of the crop at this stage and the status of the soil water salinity at deeper depths in the rootzone.

In early stages of plant development it is often desirable to irrigate just sufficiently to bring the soil to "field capacity" to the depth of present-rooting or just beyond. Eventually, however, excess water must be applied to leach salts accumulated in the upper profile to deeper depth in order to provide the growing plant access to more "usable" soil water in accordance with its expanding needs. Thus, the amount of irrigation water required is dictated by the plant's need for water, the volume of soil reservoir in need of replenishment and the level of soil salinity in the lower rootzone. Benefits of different amounts of saline irrigation water should be determined by evaluating their effects on relative crop yield using the water production function model.

5

Organic Farming

Organic farming is a form of agriculture that relies on ecosystem management and attempts to reduce or eliminate external agricultural inputs, especially synthetic ones. It is a holistic production management system that promotes and enhances agro-ecosystem health, including biodiversity, biological cycles, and soil biological activity.

In preference to the use of off-farm inputs, organic farming emphasises management practices, taking into account that regional conditions require locally adapted systems. Utilising both traditional and scientific knowledge, organic agricultural systems rely on agronomic, biological, and mechanical methods, as opposed to using synthetic materials, to fulfill any specific function within the system. Organic farming is also associated with support for principles beyond cultural practices, such as fair trade and environmental stewardship, although this does not apply to all organic farms and farmers.

Organic farming excludes the use of synthetic inputs, such as synthetic fertilisers and pesticides, and genetically modified organisms (GMOs). In many countries the use of veterinary drugs is excluded. In a number of countries, including the US, Bulgaria, Iceland, Norway, Romania, Switzerland, Turkey, Australia, India, Japan, the

Philippines, Korea, Taiwan, Thailand, Argentina, Costa Rica, Tunisia, and in the EU, organic farming is also defined by law, so that the commercial use of the term organic to describe farming and food products is regulated by the government. Where laws exist, organic certification is available to farms for a fee, and it is usually illegal for a non-certified farm to call itself or its products organic. Elsewhere, for example, in Canada, voluntary certification is available, while legislation may be pending.

Methods of organic farming vary. However, organic approaches share common goals and practices. In addition to the exclusion of synthetic agrichemicals, these include protection of the soil, promotion of biodiversity, and outdoor grazing for livestock and poultry. Within this framework, individual farmers develop their own organic production systems, determined by factors such as climate, market conditions, and local agricultural regulations.

It is important to make the distinction between organic farming and organic food. Farming is concerned with producing fresh products—vegetables, fruits, meat, dairy, eggs—for immediate consumption, or for use as ingredients in processed food. The manufacture of most commercially processed food is well beyond the scope of farming. It is also important to note that organic farming is a reaction against the large-scale, chemical-based farming practices that have become the norm in food production over the last 80 years. The differences between organic farming and modern conventional farming account for most of the controversy and claims surrounding organic agriculture and organic food.

The development of modern organic farming techniques is also a function of economics. Most of the agricultural research over the last century has concentrated on chemical-based methods— little funding and effort have been put into using current scientific tools to understand and advance organic agricultural approaches. Principles of plant cultivation, in many situations identical to those of organic farming, are applied—usually, though not necessarily, at a smaller scale—in the practice of organic horticulture.

History of Organic Farming

The organic movement began as a reaction of insiders against the industrialisation of agriculture. For some time it remained below the awareness of the food buyer. As the contrasts between organics and the new conventional agriculture grew, so to did public awareness of organic farming. This led to a distinct organic market, and, eventually, a grassroots consumer cause. Advances in biochemistry, (nitrogen fertiliser) and engineering (the internal combustion engine) in the early 20th century led to profound changes in farming. Research in plant breeding produced hybrid seeds. Fields grew in size and cropping became specialised to make efficient use of machinery and reap the benefits of the so-called "green revolution."

While some indigenous cultures had been farming organically for centuries, organic agriculture began to develop consciously in Central Europe and India in the early twentieth century as a reaction to industrialisation. The British botanist, Sir Albert Howard often called "the father of modern organic agriculture" studied traditional farming practices in Bengal, India. He came to regard such practices as superior to modern agricultural science and recorded them in his 1940 book, An Agricultural Testament.

In Germany, Rudolf Steiner's Spiritual Foundations for the Renewal of Agriculture, published in 1924, led to the popularisation of biodynamic agriculture, one of the first organic farming systems. In 1939 Lady Eve Balfour, influenced by Sir Howard's work, launched the first scientific, side-by-side comparison of organic and conventional farming in England. Called the Haughley Experiment, it was documented by Lady Balfour in her book, The Living Soil. Its influence led to the formation of the Soil Association in 1946, a key international organic advocacy group.

The first use of the term organic farming is usually credited to Lord Northbourne, in his book, Look to the Land, wherein he described a holistic, ecologically balanced approach to farming. Technological advances during World War II spurred on post-war innovation in all aspects of agriculture, resulting in such advances as large-scale irrigation, fertilisation, and the use of pesticides. Ammonium nitrate, used in munitions, became an abundantly cheap source of nitrogen. DDT, originally developed by the military to control disease-carrying insects among troops, was applied to crops, launching the era of widespread pesticide use.

During the 1950s, sustainable agriculture became a topic of scientific interest, although research focused on developing the new chemical approaches. In the US, J.I. Rodale popularised organic gardening among consumers.

In the 1970s, global movements concerned with the environment championed organic farming. As the distinction between organic and conventional food became clearer, one goal of the organic movement was to encourage consumption of locally grown food, which was promoted through slogans like "Know Your Farmer, Know Your Food". In 1972, the International Federation of Organic Agriculture Movements (IFOAM), was founded in Versailles, France. IFOAM was dedicated to the diffusion of information on the principles and practices of organic agriculture across national and linguistic boundaries.

In the 1980s, various farming and consumer groups worldwide began pressing for government regulation of organic production. This led to legislation and certification standards being enacted beginning in the 1990s. Since the early 1990s, the retail market for organic farming in developed economies has grown about 20 per cent annually due to increasing consumer demand. While small independent producers and consumers initially drove the rise of organic farming, increasingly organic market growth has led to the participation of agribusiness interests. As the volume and variety of "organic" products grows, production is increasingly large-scale.

Principles of Organic farming

There are several compelling principles that characterise certified organic farming. They include biodiversity, integration, sustainability, natural plant nutrition, natural pest management, and integrity. Most organic operations will reflect all of these to a greater or lesser degree. Since each farm is a distinct entity, there is a large degree of variation.

Biodiversity

As a general rule, diverse ecosystems in nature have a higher degree of stability than those with only a few species. The same is essentially true for agroecosystems. Farms with a diverse mix of crops have a better chance of supporting beneficial organisms that assist in pollination and pest management. Diversity above ground also suggests diversity in the soil, providing better nutrient cycling, disease suppression, tilth, and nitrogen fixation. Good organic farmers mimic the biodiversity of nature through practices like intercropping, companion planting, establishment of beneficial habitats, and crop rotation (sometimes referred to as companion planting across time). The effort to increase biodiversity works hand-in-hand with enterprise diversity, which is often (but not necessarily) an objective on organic farms.

Diversification and Integration of Enterprises

The drive to build biodiversity in organic systems encourages diversity among enterprises, but not as isolated or independent entities. Good organic operations integrate their various enterprises. A good example can be seen among Midwestern organic crop and livestock operations.

The typical organic operation ties the needs of crops and livestock together in a practi-cal and elegant way. The forage and grain needs of ruminant livestock make for a diverse mix of crops. Particularly valuable is the inclusion of le-gume forages for ruminant feed. Forage legumes in rotation fix a sustainable supply of nitrogen in the soil that feeds subsequent non-legume crops in rotation. Manure from the livestock enterprises is conserved as a nutrient resource and recycled back to the crop fields. Farms such as these have the additional advantage of greater economic sustainability, as their risks are spread over several livestock and crop enterprises.

Sustainability

In addition to the greater economic sustainability afforded by enterprise diversification, organic farmers are often able to reap market premiums for certified production. However, since many organ-ic enterprises realise somewhat lower marketable yields, this has not always translated into higher profits or greater economic sustainability. As more and more organic growers enter the marketplace, it is likely that premiums will stabilise at modest levels and may vanish for some crops. Organic producers need to look well ahead and be aware of shifting trends.

As alluded to earlier, many organic farms per-form well on many of the measurable indicators associated with sustainability, such as energy con-sumption and environmental protection. However, sustainability is an ideal, and the best that can be said is that current organic farms are closer to the ideal than most alternatives—certainly closer than comparable conventional farming operations.

The extent to which traditional organic agriculture philosophy influences the adoption of sustainable practices has only been touched upon. For example, during the Washington University studies of midwestern farms in the late 1970s, researchers observed that organic farmers had embraced conservation tillage technologies at a much faster rate than their conventional counterparts. Conservation tillage was not and is not considered a traditional practice of organic farming, yet its ready adoption points to the dynamic nature of organic agriculture and offers clear evidence that the underlying philosophy of sustainability—strongly championed by Albert Howard—remains a vital part of organics. Given the option of a sustainable technology that fits the constraints of certified organic agriculture, it is natural for most organic farmers to choose it.

Natural Plant Nutrition

Even though we require the same basic "stuff" to live, it is somewhat challenging to draw simple comparisons between the nutritional needs and processes of plants and those of animals. Plants are able to photosynthesise to make sugars, which are ultimately synthesised into proteins and other plant constituents. Humans and other animals, by contrast, can obtain energy foods, proteins, and vitamins only by consuming plants or other animals.

Both plants and animals also require minerals. Humans and other animals extract minerals, along with sugars and proteins, from the food they eat. Plants, too, obtain minerals—and a wide range of vitamins, antibiotics, and other useful compounds—through digestion. However, plant digestive systems are not internalised as they are in animals. Plants must rely on the external digestive processes of the soil system within reach of their roots—a zone called the rhizosphere.

Nutrient Absorption

Critics are often under the illusion that organic farmers believe plants obtain all their nutrients from an organically managed soil in a chemically organic form. While a few organicists may believe that, the majority recognises that digestion processes in the soil release minerals in forms similar to those applied as commercial fertilisers. Unfortunately, the notion that organic farmers are naïve and ignorant about basic agronomy is a red herring that has often foiled intelligent discussion about the pros and cons of the system. Among the facts that are often obscured is that plants can and do absorb significant amounts of large organic molecules from the soil; herbicide and systemic insecticides are among these. In healthy soils they also absorb vitamins, chelated minerals, hormones, and other beneficial compounds.

The organic philosophy of crop nutrition begins with proper care and nourishment of the organisms responsible for the soil digestive process. Organic farmers believe this is best accomplished by avoiding toxic chemicals and practices—like excessive tillage—that are harmful to soil organisms, as well as by the addition of organic matter and natural rock minerals. Conventional systems, in contrast, try to circumvent the soil's digestive process and provide needed minerals to the plant directly, in a soluble form.

From the organic perspective, the conventional approach has several flaws.

- Applying large quantities of soluble fertiliser to a crop only one, two, or three times per season floods the plant with those nutrients, causing nutritional imbalances that lead to crop diseases, insect infestations, and reduced food quality.
- Failure to support and care for soil biotic life, along with other practices that are downright destructive, ultimately leads to its decline. As a result, plants lose out on the vitamins and other beneficial products these organisms produce, tilth is reduced, and the soil becomes increasingly dependent on synthetic inputs.
- Conventional fertilisation tends to concentrate on a limited number of macronutrients, even though the need for at least 13 soil minerals is scientifically recognised. This skewed focus is also responsible for generating imbalances in the plant.
- Application of large àmounts of soluble nutrients can stimulate certain problem weed species.
- Soluble nutrients—especially nitrate—are prone to leaching, which can cause a number of environmental and health problems.It is organic farming's approach to soil building and plant fertilisation that is the true basis for the belief that organic food and feed has superior nutritional value, much more so than the absence of pesticide residues, which has drawn the spotlight ever since the 1960s.

Natural Pest Management

Whether conventional or organic, all farmers are concerned with pests. They spend a lot of time and resources controlling them. However, in the organic "world view," pests—whether weeds, insects or diseases—are not simply scourges. They are indicators of how far a production system has strayed from the natural ecosystems it should imitate. Certain weeds, for example, tend to predominate when soils are too acidic or too basic; some become a problem when soil structure is poor and conditions become anaerobic; others may be stimulated by excessive fertiliser or manure salts.

Organic proponents also believe that insect pests are attracted to inferior or weak plants—the result of poor crop nutrition. Their logic continues by asserting that pests are naturally repelled by vigorous, well-nourished plants. This belief is often challenged, and significant research remains to be done.

As scientific understanding has grown, insect pest outbreaks are also being understood as imbalances in the whole agroecosystem and how it is managed. In nature, massive pest outbreaks are relatively rare and short-lived, due to the presence of natural predators, parasites, and disease agents that quickly knock the pest numbers back down to a moderate level. In farming systems that inadvertently destroy or otherwise fail to support the natural control complex, pest problems are routine and, typically, worsen with time. The farmer becomes increasingly addicted to costly and extreme control methods to produce a crop. Most organic growers consider pesticides to be a cause of agroecosystem imbalances and employ allowed natural pesticides as little as possible.

Integrity

Integrity refers to the systems in place and actions undertaken to assure that consumers of organic products get what they pay for. Consumers have a right to expect that the organic food they buy not only be raised by organic methods but be protected from contamination and from commingling with non-organic products.

While the responsibility for much of this rests with others in the organic marketing chain, many certified organic growers need to incorporate additional practices that work to assure the integrity of their products. Proper record keeping is very important in this regard, though growers are often reluctant to spend much time on it. Among the more important production practices in the field are buffer strips, which reduce chemical drift from neighboring fields and roadsides, while also serving water and soil conservation objectives.

Tools and Practices

The tools and practices of organic agriculture include traditional alternatives—crop rotation, manuring, liming, etc.—long recognised as important to a sound production system. They also include more contemporary practices and materials that research and keen observation over time have contributed. The following list of tools and practices is not intended to be comprehensive, though the primary options are addressed. Note, too, that each farm operation will employ its own combination of tools and practices to build a working organic system. There is no simple cookbook formula for combining them in ideal proportions.

Planned Crop Rotation

Essentially a tool for annual cropping systems, crop

rotation refers to the sequence of crops and cover crops grown on a specific field. Particular sequences confer particular benefits to long and short-term soil fertility, and to pest management.

Agronomic operations are especially dependent on crop rotations that include forage legumes. These provide the vast majority of the nitrogen required by subsequent crops like corn, which is a heavy consumer of that nutrient. Even when livestock are present to generate manure, the animals are largely recycling the nitrogen originally fixed by legumes in the system.

The basic rotation demonstrates the elegant way in which a whole farm system can function:

- Legumes fix nitrogen in the soil, providing for subsequent non-legumes in the rotation.
- Several insect pest cycles are interrupted, especially those of the northern and western rootworm species that can be devastating to corn.
- Several plant diseases are suppressed, in cluding soybean cyst nematode.
- Weed control is enhanced as perennial weeds are destroyed through cultivation of annual grains; most annual weeds are smothered or eliminated by mowing when alfalfa is in production.
- Livestock manures are applied just in advance of corn, a heavy nitrogen consumer.
- All crops can be marketed either as is or fed to livestock to be converted into value- added milk, meat or other livestock products.

Green Manures and Cover Crops

Green manuring consists of incorporating into the soil a crop grown for the purposes of soil improvement. It is a practice with a long history. Green manuring has been ignored in recent years as a serious option for soil improvement because the traditional practice entailed planting a full-season cover crop. This removed the field from commercial production for a whole season. Interest has returned, however, since green manuring strategies have been combined with cover cropping schemes.

Cover cropping is growing a crop for the purpose of soil and nutrient conservation. It is a more contemporary concept than green manuring, in crop agriculture. The two concepts—cover cropping and green manuring—go well together, as most cover crops are easily used as green manures prior to the planting of a commercial crop. The combined benefits become economically feasible when the cover is grown during the off-season or inter-seeded with the main crop. It is made even more desirable when the cover crop includes nitrogen-fixing legumes.

Manuring and Composting

Livestock manures are the most traditional and widely recognised organic fertilisers. Under ideal circumstances, livestock enterprises are integrated into the whole farm operation, and manuring becomes part of a closed system of nutrient recycling. This is still strongly encouraged in organic operations. In reality, however, crops and livestock production are often divorced from each other, and manures must be imported.

This has created some concerns in the organic community, since much manure is now generated by large, industrial agriculture feeding operations called CAFOs (Confined Animal Feeding Operations). Not only are there concerns about contaminants (heavy metals, antibiotics, pesticides, hormones) but many in the organic community also object to any "partnering" with this segment of conventional agriculture, which is considered at odds with the environmental and social values represented by organic farming.

Nonetheless, the National Organic Program does not differentiate between CAFO and other livestock manure sources. However, the NOP regulations do require that livestock manure not contain any synthetic substances not included on the National List of synthetic substances allowed for use in organic crop production.

Another issue that has grown up around manure use in organic farming relates to food safety. At a time when concerns about microbial contamination are high, there are questions about the risks associated with manure use on food crops. A focus piece on the February 2000 television news program 20/20 was especially controversial. The segment suggested that organic foods were more dangerous than other food products in the marketplace due to manure fertilisation. The reporter ignored the fact that conventional farms also use manures. Were all the manure generated annually in the U.S. applied only to organic farm acreage, each acre would receive about 933 tons. Furthermore, certified organic producers have strict guidelines to follow in handling and applying manures.

The National Organic Program regulations require raw animal manure be incorporated into the soil not less than 120 days prior to the harvest of a product whose edible portion has direct contact with the soil surface or soil particles, and be incorporated into the soil not less than 90 days prior to harvest of a product whose edible portion does not have direct contact with the soil surface or soil particles.

One of the best means of handling manures is composting. Composting stabilises the nutrients in manure, builds populations of beneficial organisms, and has a highly beneficial effect on soils and crops. Compost can be produced on-farm by a number of means. Additional products from composts, such as compost teas, have special applications in organic agriculture.

Livestock on Organic Farms

Among the thorniest of issues swirling around the edges of organic agriculture is the role of livestock. The disagreements arise because of the diversity of people and philosophies in the organic community. Organic agriculture can usually count vegetarians and animal welfare proponents among its more vocal supporters. Many of these people feel strongly that animals should not be exploited. Their rationale often goes beyond emotional and religious beliefs; convincing human health concerns, social issues, and environmental reasons are commonly cited. On the other side of this argument are those who feel that an organic farm cannot achieve its full potential or ecological balance without livestock manure; that it is essential to nutrient cycling and to the finer aspects of soil building.

Excellent soil fertility can be built in the absence of farm livestock and livestock manures by using vegetation-based composts and by harnessing the livestock in the soil—earthworms and other soil organisms. However, it is clearly easier to design a contemporary, low-input organic farm when traditional livestock are integrated. The biological and enterprise diversity that livestock can bring contributes enormously to stability and sustainability. A good example is provided by Rivendell Gardens in Arkansas, which began integrating livestock enterprises after several years as a solely horticultural operation.

The owners of Rivendell, Gordon and Susan Watkins, now rotate their strawberry and vegetable crops with grass-fed, direct-marketed beef and pastured poultry. Ideally, poultry follows beef on pasture to reduce cattle parasites. The seasons in mixed legume/grass pasture leave the soil quite mellow and well-manured for subsequent high-dollar horticultural crops.

The Rivendell operation demonstrates the sort of organic management where a large number of organic farmers and many animal welfare proponents find common ground. The Watkins' animals are all raised with minimal confinement and generous access to sunshine, fresh air, and free-choice foodstuffs. While domesticated and destined for slaughter, they lead low-stress lives in conditions much closer to natural than the conventional alternatives. This is the antithesis of industrialised factory farming systems, which are increasingly becoming the norm in livestock production.

Many in both the organic and animal welfare communities are working to prohibit factory farming of livestock in organic systems. Many of the difficulties revolve around the fine interpretations of language in various organic standards. Wording such as "access to fresh air and sunlight," for example, can be construed to mean nothing more than opening the door on one end of a large confinement poultry house for a couple hours a day.

Intercropping and Companion Planting

Interplanting two or more mutually beneficial crops in close proximity is one strategy for increasing biodiversity. In large-scale mechanised crop culture, this is called intercropping. It typically involves alternating rows or a number of rows of compatible field crops, like soybeans and corn. It also applies to sowing multiple forage crops, like alfalfa, bromegrass, and timothy, when these are grown together.

When interplanting is done on a smaller scale, it is often called companion planting. A classic example of companion planting is the inter-planting of corn with pole beans and vining squash or pumpkins. In this system, the beans provide nitrogen; the corn provides support for the beans and a "screen" against squash vine borer; the vining squash provides a weed suppressive canopy and discourages corn-hungry raccoons.

Biological Pest Control

Organic farming relies heavily on populations of beneficial insect predators and parasites, pest disease agents, insect-eating birds and bats, and other creatures, to help manage pest problems. These biological controls help keep pest numbers at levels where further cultural activities or relatively mild pesticides are adequate to assure a crop. In some instances, biological control can be so effective that no additional action is even needed by the farmer.

Some see biological control as a default benefit of the soil fertility practices of organic farming. The diversity of crops in a soil-building rotation, the use of cover crops, and other practices build a diverse soil biology that works to keep soil pests in check. They also provide substantial above-ground habitat for beneficials. The absence of pesticides also favors biocontrol.

In many organic systems, farmers sometimes purchase and release control agents like ladybird beetles, lacewings, trichogramma wasps, etc., or use weeder geese—a quaint but effective biological weed control.

Increasingly, growers are designing and maintaining both permanent and temporary habitats specifically for beneficial insects, spiders, and other helpful species. This is known as farmscaping.

Sanitation

Sanitation can take on many forms:

- — removal, burning, or deep plowing of crop residues that could carry plant disease or insect pest agents
- — destruction of nearby weedy habitats that shelter pests
- — cleaning accumulated weed seeds from farm equipment before entering a new, "clean" field
- — sterilising pruning tools

As in human and animal health, sanitation practices can go a long way in preventing crop pest problems. However, many practices—such as clean cultivation, deep plowing, and burning crop residues—can increase erosion and reduce biodiversity. Thus, they may conflict with sustainability. Good organic growers recognise this and treat those practices as transitional or rescue options, rather than relying on them on an annual basis.

Tillage and Cultivation

Tillage and cultivation are tools that can accomplish a variety of objectives in farming systems: weed control, crop residue management, soil aeration, conservation of manures and other fertilisers, hardpan reduction, sanitation to destroy pest and disease habitat, etc.

While conventional farmers rely on chemicals to accomplish many of these objectives, organic growers focus more on improving tillage and maximising its benefits. Guidelines for primary tillage, for example, are intent on conserving crop residues and added manures in the upper, biologically active zones of the soil, rather than burying them deeply where decomposition is anaerobic. Leaving soils completely bare and vulnerable to erosion is discouraged; fall moldboard plowing is certainly frowned upon.

Cultivation in organic systems often rises to the level of art. Row-crop farmers frequently use blind cultivation -shallow tillage, which largely ignores the crop rows-beginning shortly after seeding until the plants are but a few inches high. Rotary hoes, wire-tooth harrows, and similar equipment can be used for blind cultivation, delaying the first flush of weeds and giving the crop a head start.

Conservation Tillage and Organic Farming

Organic agriculture is often characterised as addicted to maximum tillage—with growers using every opportunity to lay the land bare with shovel, plow, or rototiller. This image has been magnified through the popularity of small-scale organic systems like the French Intensive and Biointensive Mini Farming models that espouse double- and triple-digging to create deep rooting beds. While appropriate to such intensive systems, this degree of cultivation is not characteristic of organic agriculture in general. It may surprise some to learn that a large number of organic producers are not only interested in conservation tillage, they have adopted it. This will be a surprise because many believe that conservation tillage always requires herbicides.

The interest in conservation tillage among organic producers in the Cornbelt was well documented in the mid-1970s by Washington University researchers. They noted that the vast majority of organic farmers participating in their studies had abandoned the moldboard plow for chisel plows. Plowing with a chisel implement is a form of mulch tillage, in which residues are mixed in the upper layers of the soil, and a significant percentage remains on the soil surface to reduce erosion. Furthermore, a notable number of organic farmers had gone further to adopt ridge-tillage, a system with even greater potential to reduce erosion. It was especially interesting to note that the use of these conservation technologies was almost nil among neighboring conventional farms at the time. Organic growers were actually pioneers of conservation tillage in their communities.

After blind cultivation, subsequent weed control operations in larger-scale systems can make use of advances in tillage equipment such as rolling cultivators, finger weeders, and torsion weeders that allow tilling close to the plant row. Smaller-scale operations often use wheel hoes, stirrup hoes, and other less capital-intensive hardware. Determining the amount, the timing, and the kind of tillage to be done can be a balancing act for the organic grower, but experience and observation over time lead to proficiency.

There are downsides to tillage, however, and most organic growers are well aware of them. The most obvious of these is the dollar cost; organic farmers are as concerned as their conventional counterparts about costs of production and strive to minimise expensive field operations. There is also a cost to the soil and environment. Every tillage operation aerates the soil and speeds the decomposition of the organic fraction. While this may provide a boost to the current crop, it can be overdone and "burn up" the humus reserves in the soil. Excessive tillage can also be directly destructive to earthworms and their tunneling, reducing their benefits to the land. There is also the danger of compaction, even when field operations are well timed.

Mulching

Mulching is a practice often used by organic growers. Traditionally, it entails the spreading of large amounts of organic materials—straw, old hay, wood chips, etc.—over otherwise bare soil between and among crop plants. Organic mulches regulate soil moisture and temperature, suppress weeds, and provide organic matter to the soil. Mulching is most appropriate to small, intensive operations with high-value annual or fruit crops.

A few systems of no-till organic gardening have evolved from the concept of deep, permanent mulching. Plastic mulch, as long as it is removed at end of growing or harvest season, is also permitted in certified organic production. Its use allows larger acreage to be brought more easily under herbicide-free management, though there are serious issues to be addressed.

High-Input Organic Agriculture

Certified organic farming can entail much greater reliance on off-farm inputs. Intensive annual strawberry and vegetable systems under plasticulture are good examples. In these systems, traditional rotations and soil building practices are usually employed, followed by clean cultivation and the laying of plastic mulch and drip irrigation tape on shaped beds. During the season, large amounts of soluble organic fertilisers—typically fish-based—are fed to the crop through the drip system. At the end of the season, all plastics must be removed from the field, and it is returned to more standard organic management. Ideally, an off-season cover crop will be planted. Such systems are often exceptionally productive and economically attractive, when organic premiums are good. The high cost of soluble organic fertiliser, however, plus the marginally higher cost of pest controls, make such systems largely non-competitive in the conventional marketplace.

The labeling of such high-input systems as organic presents a paradox for many proponents of organic agriculture. It is unclear whether these technological advancements reflect the kind of farming most practitioners and supporters of organics think of as truly "organic." To begin with, the research citing environmental and economic benefits has largely been done on low-input organic systems; it is questionable whether similar findings would be made about high-input systems, especially regarding environmental matters.

Of particular note, while low-input organic systems are documented as being more resistant to erosion, fields under plastic mulch are reported to be fifteen times more erodible. Traditional organic farms leach minimal amounts of nitrogen into tile or groundwater; the losses from fields loaded with high levels of soluble organic fertilisers is certain to be greater, but how much greater is unknown. The fossil fuel energy involved in plastic manufacture, transportation, and application may or may not be compensated by reductions in tractor fuel use.

Finally, the lowered capital investment required to produce a crop by traditional organic methods makes this form of farming more accessible to resource-poor farmers and entails less risk in years of crop failure or lack of premiums. These factors are less certain in a high-input system. A further consideration is the issue of plastic disposal following removal. At this time, there are few to no options for recycling, and landfills are the fate of plastic mulches at the end of each season. While it is unwise to rush to judgement regarding high-input organic farming, it is clear that some adaptations will need to be made, if the traditional character and sustainability benefits of organic farming are to be preserved.

Fire

While fire can be used in a number of ways in organic agriculture, the area of greatest interest is flame or thermal weeding. In its most common application, torches mounted on a tractor toolbar throw a hot flame at the base of mature plants, over the inter-row area, or both. Tractor speed is adjusted so that weeds are not burned so much as seared. Searing is sufficient to kill most seedling weeds and uses less fuel. Liquid propane (LP) gas is the fuel most commonly used, though alternatives such as alcohol and methane offer the possibility of on-farm sources.

Supplemental Fertilisation

In many organic systems, crop rotation, manuring, green manuring, along with enhanced biological activity in the soil, provide an abundant supply of plant-essential minerals annually. This is especially true on naturally deep and rich prairie soils. It is less true on poorer soils and on those that have been heavily exploited through non-sustainable farming practices. To correct mineral deficiencies in organically managed soils, organic growers often apply ground or powdered rock minerals.

The most commonly used rock mineral is high-calcium aglime. Dolomitic limestone, various rock phosphates, gypsum, sulfate of potash-magnesia, and mined potassium sulfate are also common. These are all significant sources of primary (P, K) and/or secondary (Ca, Mg, S) plant nutrients. The savvy organic grower applies significant amounts of these materials only with the guidance of regular soil testing.

Less common are other rock powders and fines that are limited sources for the major nutrients but are rich in micronutrients or have some other soil-improving characteristic. Among these are glauconite (greensand), glacial gravel dust, lava sand, Azomite, granite meal, and others.

Supplementary nutrients that include nitrogen are often provided in the form of animal or plant products and by-products such as fish emulsion, blood meal, feather meal, bone meal, alfalfa meal, and soybean meal. Most of these products also supply some organic matter, though that is not the primary reason they are used.

Evaluating Tools and Practices

One basis on which to evaluate the tools or practices one chooses for an organic operation is whether or not they contribute to biodiversity—biodiversity being one of the principal characteristics of a sustainable organic agriculture.

Crop rotation, cover cropping, farmscaping, companion planting, and intercropping are outstanding examples of practices that contribute to biodiversity. They therefore contribute to the long-term stability and sustainability of the farm agroecosystem. Composting and manuring likewise contribute to biodiversity, but since the diversity they promote is mostly in soil biota, it is rather less obvious to the casual observer.

On the other end of the spectrum are practices such as tillage, cultivation, thermal weeding, solarisation, and plastic mulching. Such tools significantly reduce diversity in the field and tend to move the system in a less sustainable direction. This, however, does not necessarily make these practices bad choices.

Organic farming has often been called natural farming, as it tries to mimic the processes of nature in producing crops and livestock. However, the analogy goes only so far, since most agricultural systems are characterised by a struggle between human and nature, each with a clear notion of what plants and animals the land ought to support and in what proportions. The farmer is typically in the position of "holding back" the natural succession of plant and animal species through the use of diversity-reducing tools and practices. The ideal is to bring about an agricultural system in which the long-term direction emphasises diversity and sustainability. Among the best visualisations of this ideal are those emerging from the Permaculture movement.

Balanced Nutrition

Whether or not it has been customary in the past, organic growers are encouraged to have periodic soil testing done on their fields. How the results of a soil test are used, however, can vary considerably among farmers, depending on their personal philosophy and management skill.

While there is certainly a segment of the organic farming community that has no faith in soil audits, most growers use them as a means to monitor progress in building their soils, to identify nutrient deficits, and to guide supplementary fertilisation. While there are many ideas about fertilisation guidelines, there are two schools of thought that dominate.

The first is commonly known as the sufficiency approach or model. Though somewhat oversimplified, the following are among its significant characteristics:

— annual fertiliser additions of P and K are based in good part on how much the crop is expected to remove from the soil at harvest

— additional amounts of P and K are recommended based on keeping the soil nutrient reserves at a particular level

— lime is added to the soil based on pH

— little to no attention is paid to nutrient balance or to the levels of secondary nutrients Ca and Mg.

The second approach is referred to variously as cation nutrient balancing, the Albrecht system, the CEC, or the base saturation approach. It differs from the sufficiency approach in that fertiliser and lime recommendations are made based on an idealised ratio of nutrients in the soil and its capacity to hold those nutrients against leaching.

Cation nutrient balancing is more popular among practitioners of organic farming and sustainable agriculture in general than it is among conventional growers. However, there is no universal agreement on which approach is most appropriate for organic management.

A wide range of other products -humates, humic acids, enzymes, catalyst waters, bioactivators, surfactants, to name a few—are also acceptable in organic crop production. However, the tradeoff between out-of-pocket cost and efficacy of such materials is often challenged by conventional and organic growers alike. Organic growers are encouraged to experiment, but to do so in a manner that allows the actual results to be measured.

Biorational Pesticides

While, in principle, any pesticide use is discouraged in organic systems, a rather wide range of biorational pesticides is permitted. The frequency of pesticide use varies considerably with crop and location. For example, there is virtually no use of pesticides on organic row crops in the Cornbelt. By contrast, organic tree fruits in the Midsouth routinely receive heavy applications of several fungicides and insecticides allowable in organic production.

The pesticides permitted in organic farming fall predominantly into several classes.

— *Minerals*: These include sulfur, copper, diatomaceous earth, and clay-based materials.

— *Botanicals*: Botanicals include common commercial materials such as rotenone, neem, and pyrethrum. Less common botanicals include quassia, equisetum, and ryania. Tobacco products are also botanicals but are prohibited in organic production due to their high toxicity.

— *Soaps:* A number of commercial soap-based products are effective as insecticides, herbicides, fungicides, and algicides. Detergent-based products are not allowed for crop use in organic production.

— *Pheromones:* Pheromones can be used as a means to confuse and disrupt pests during their mating cycles, or to draw them into traps.

— *Biologicals:* One of the fastest-growing areas in pesticide development, biopesticides present some of the greatest hope for organic control of highly destructive pests. Among the most well-known biopesticides are the Bacillus thuringiensis (Bt) formulations for control of lepidopterous pests and Colorado potato beetle.

Use in Organic Crop Production

One of the greater difficulties that organic producers face on a regular basis is determining whether or not a particular product or material can be used in organic production. Sad to say, the problems are real, but some basic clarifications will help. First of all, all natural or nonsynthetic materials can be assumed to be acceprable in organic production. There are a few exceptions, however, which will be explained shortly.

Seeds, Seedlings and GMOs

One of the challenges faced by organic growers is getting suitable planting stock. Certified production requires that seed not be treated with pesticides and be organically grown.

Transplants must be purchased from a certified organic source or otherwise be grown using organic methods. This presents some difficulty, as most commercial potting mixes contain prohibited fertilisers and wetting agents, requiring special ordering or the added effort of making a homemade mix.

Generally, planting stock for perennial crops like tree fruits and berries is available from conventional sources, even if treated with pesticides. However, no production can be sold as organic for a minimum of 12 months following transplanting to an organic field.

In all instances, organic growers are not permitted to use varieties of crops that have been developed through genetic engineering (GE). At first glance, this might seem perplexing. GE crops promise further means of non-chemical pest control and the possibility of nutritionally enhanced foods. However, the organic community is concerned about environmental, economic, and social impacts from this new technology which, they feel, have not been adequately studied. This same concern applies to GE foods. Natural foods consumers—a large segment of the organic market—do not want them. The backlash was so strong that not only GE crop varieties, but all inputs, such as GMO-derived biopesticides, are prohibited from organic production.

Unfortunately the GE issue does not resolve itself so simply for organic farmers. The proliferation of GMOs in the marketplace and across the landscape has created a host of new challenges that have particular implications for organic agriculture:

— *Genetic drift onto crops.* Pollen drifting from adjacent fields can contaminate organic crops. Some clients in the lucrative European Union market accept 0% GMO contamination; others accept only very low levels. Pollen from cross-pollinating crops like corn easily travels hundreds of yards and may be carried for miles under the right conditions. No reasonable amount of non-crop buffer can prevent contamination except in highly isolated locations.

— *Genetic drift onto soils.* Concerns have been expressed over the persistence of GE pollen in soils.

— *Non-labeled products.* Should soybean, cottonseed, and corn gluten meals be prohibited as fertilisers, since these may come from GE varieties? Should pesticidal use of vegetable oils be prohibited?

— *Pest control problems.* The rapid proliferation of corn, cotton, and other crops featuring the Bt gene from Bacillus thuringiensis is likely to accelerate the development of resistance to this natural pesticide. Organic growers may lose one of their most useful pest management tools.

— *Domination of the food system.* In the late 1990s, the public was made aware of two specific developments of agricultural biotechnology dubbed the terminator and traitor technologies. Terminator technology is the genetic altering of seed so that the subsequent generation will not germinate. This prevents farmers from saving a portion of the harvested crop to replant the next year—a traditional practice among farmers worldwide. Traitor or suicide technology is the genetic altering of seed so that it will fail to germinate or mature unless a proprietary chemical is applied. Such technologies are authoritarian and advance the centralisation and industrialisation of the food system. Organic farmers perceive additional challenges from genetic drift impacts and difficulty in finding non-GE seed in the marketplace.

Foliar Fertilisation

Foliar fertilisation or feeding entails the application—via spraying—of nutrients to plant leaves and stems and their absorption at those sites. It is not specifically an organic practice, though it is commonly used by many organic growers. The fertiliser materials used are typically soluble fish- and seaweed-based products, naturally chelated nutrients, humic acid extracts, and teas made from plants, dried blood, manure, guano, or compost.

At first glance, the use of foliar feeding appears contradictory to the organic notion that one feeds the soil to feed the plant. Organic growers rationalise the use of this approach on two points. Foliar feeding is strictly supplemental fertilisation; it is not used as a substitute for traditional soil building practices. Foliar fertilisation is understood to increase the production of root exudates, which stimulates biological activity in the rhisosphere. The soil bio-life gets considerable benefit in this indirect way from foliar feeding.

Esoteric Practices

There are a number of farming and gardening practices based on belief in a non-physical world closely aligned with our physical reality. Those who use these practices believe that conditions in this unseen realm influence, or even dictate, what happens on a material level. They take pains to understand these influences and adjust their farm activities. In some instances, they act to influence events. Obviously, these practices are rooted more deeply in metaphysics than in the conventional sciences.

The most well-known of these esoteric practices is the scheduling of planting or other field operations according to lunar or other astrological signs. Dowsing and other forms of divining may also be used to guide scheduling, fertiliser selection, and other facets of farming. The use of potentised preparations, as done in Biodynamic farming, is also a metaphysical practice. Radionics, energy balancing towers, "medicine" wheels, prayer—these are further examples of practices used by some farmers to bring a tangible spiritual element into their farm operation.

Esoteric practices like these are not inherently part of organic agriculture. In all likelihood, only a small-to-modest minority practice them or believe in their efficacy. Still, esoteric practices are sometimes associated with organic agriculture, and it is almost a given that practitioners of "energetic agriculture" are either certified organic or actively pursue a sustainable form of agriculture.

The rationale for this association of organic with the esoteric is not difficult to understand. Most metaphysical practices are founded on a world-view similar to that of ecology, in which everything is related to everything else. The difference is that ecology considers this interrelationship only on a hard physical level; metaphysics perceives it also on a non-physical level.

This belief tends to cast humankind as part of nature, not merely as an inhabitant. This perspective would tend to make one more concerned about the environment and, therefore, more readily drawn to organics as a philosophy and pragmatic farming approach. To put it in a nutshell, it is natural for the metaphysically-inclined to choose organics; it is far less likely, however, that the average organic farmer is drawn to metaphysical practices.

A further factor plays into the association between esoteric farming practices and organics. It relates to the fact that organic agriculture has, for many years, been marginalised and treated as pseudoscience by the mainstream—an experience shared by practitioners of metaphysical arts. It is not surprising for underdogs to seek common ground.

Buffers and Barriers

Field buffers are strips of sod or other permanent vegetation at the edge of or surrounding crop fields. There is considerable interest in both conventional and alternative agriculture in these buffers, since they assist in reducing soil erosion and improving water quality. If managed properly, some buffers also serve as beneficial insect habitat.

In certified organic production, however, field buffers have an added purpose in reducing crop contamination from chemicals used on adjacent land. Most agencies require a minimum 25-foot buffer along "uncontrolled" borders where there is a hazard of chemical use. Wider borders may be required where hazards are great—for instance, where adjacent farms have synthetic pesticides applied by aircraft.

Niche Market for Organic Farming

The pioneers of organic farming considered organics the preferred direction for the whole of agriculture to take. It is likely that most contemporary proponents still hold that view. While recent growth in the organic industry is definitely encouraging, much of the impetus is tied to its growth as a niche market, not as a serious shift in the direction of mainstream agriculture. Unfortunately, a likely reason for the newfound "tolerance" of organic agriculture in many land grant universities and other formerly hostile environs is its perception as a niche market opportunity only. As such, it is not a serious threat to the status quo.

Is it practical and responsible to promote organic agriculture as the dominant approach to farming in the future? While some are quick to answer, "Yes!", most are buffaloed by the nagging question, "Can organic agriculture feed the world?" Questions about the productivity and the prospects of widespread starvation have long been an effective tactic for stonewalling serious discussion of organics. While it is not the purpose of this publication to settle such a compelling issue, there are a few points that can and should be made regarding the "starving billions" scenario.

Critics warn that, were organic farming to be adopted on a wider scale, per-acre agricultural productivity would decline sharply; that meeting food needs would necessitate the plowing of even more erodible hillsides and the draining of more wetlands. Such scenarios are often based on pre-chemical era yield data that ignore the advancements of modern organic farming.

Contemporary research on organic systems—as limited as it is—indicates that the per-acre productivity of organic and conventional systems is not vastly different. Other scenarios are sometimes based on the perceived need for vast quantities of grain—particularly corn. Since overproduction of grain has led to prices below costs of production for several years, this is a questionable basis for argument. It also ignores the fact that most of the grain produced is fed to livestock and that many livestock species can eat forage instead. Ruminant livestock—cattle and sheep especially—are designed by nature to thrive on forage and can even be finished for market on pasture.

The paucity of good research on the productivity of organic systems also spotlights the lack of practical research for organic producers. That organic farms are as productive as they appear to be is remarkable in light of minimal research and extension support for many decades. It has often been argued that, had comparable resources been put into organic research, its widespread feasibility would be unquestionable. This assertion is well-supported by the development, in recent years, of a number of alternative, organically acceptable pesticides, that are certain to expand organic production of many crops in regions where it had previously been near impossible on a commercial basis.

Increasing agricultural production alone does not alleviate hunger. The amount of grain produced in the world in 1999 could, by itself, sustain 8 billion people—2 billion more than our current population. Total food production is estimated to provide each human being with at least 3500 calories per day. The issue of hunger, it appears, is not so much a lack of food, but lack of entitlement to food. People are shocked to learn that while many Bangladeshees starved during the 1974 floods, roughly 4 million tons of rice produced in that country were stacked in warehouses for want of buyers.

The people were simply too poor to buy it. They are surprised to learn that Ireland exported grain during the Irish potato famine while 1 million of its citizens died and even more emigrated; that India regularly exports food and animal feed despite an estimated 200 million in starvation. Those lacking the ready cash to buy food or the resources to produce it themselves seem destined for hunger no matter what miracles agricultural technology provides. The world's nations will need to deal with issues of equity and democracy first, if hunger is ever to be effectively addressed.

No agriculture can continue to feed a growing population if it depletes or fouls its resource base. The path undertaken by conventional agriculture is ultimately a dead end in this regard, though there is an almost mystical faith that genetic engineering and other complex technologies will always triumph. Agriculture needs to be sustainable. Therefore, those who promote organic agriculture as a true alternative are well advised to do their part in ensuring that certification and regulation does not create a "compliance agriculture" in which sustainability becomes little more than an afterthought.

Record Keeping

Few would consider record keeping as a tool for organic crop production. However, the documentation of how and where a crop was raised, what products were applied and when, which bin or cooler it was stored in, is of critical importance to establishing the integrity of the product. If the farmer cannot provide reasonable documentation that his or her crop was organically grown, that it has not been contaminated with chemicals, and that it has not commingled with a similar conventional product, then certification may well be denied.

Issues in Organic Farming

All aspects of organic farming and organic food are under debate. Environmentalists, food safety advocates, various consumer protection, social justice and labor groups, small independent farmers, and a growing number of food consumers are ranged against agribusiness and current government agricultural policies. The controversy centers on the overall value and safety of chemical agriculture, with organic farming popularly regarded as the "opposite" of modern, large-scale, chemical-based, vertically integrated, corporate food production. As public awareness increases, there are a number of obstacles to an easy grasp of the overall situation.

In recent decades, food production has moved out of the public eye. In developed nations, where most of the world's wealth, consumption, and agricultural policy-making are centered, many are unaware of how their food is produced, or even that food, like energy, is not unlimited. If the methods used to produce food are rapidly destroying the capacity for continued production, then sustainable, organic farming is as crucial a topic as renewable energy and pollution control. This proposition is at the center of most organic farming issues.

It is useful to make a distinction between organic farming and organic food. Whether organic food is tastier, safer or more nutritious has little to do with the effects of chemical agriculture on the environment. In any case, most food dollars are spent on processed food products, the manufacture of which is beyond the scope of farming. There are separate food and farming issues and lumping the two together only confuses the discussion.

The distinction between organic farming and organic certification is also important. Defining organic farming with checklists of acceptable and prohibited inputs and practices elicits similar criticisms as those leveled at chemical farming. With rules come exceptions, whether well-intentioned or purely profit-oriented, and critics hold that this can only undermine organic principles.

What is "more-or-less organic"? Certification also allows agribusiness to lobby for favorable definitions—anything that can be approved becomes "organic".

Of course, the issues, particularly the social ones, will shift if agribusiness fully adapts to and dominates organic farming, and this is the current trend. Then, large-scale, certified organic farms would probably operate much more like conventional farms do today. Environmental benefits may accrue from a change in types of pesticides and fertiliser used, more crop diversity, and the like, but if the overall agribusiness philosophy remains essentially unchanged, "organic farming" could become the norm, without any great environmental or social improvements.

Organic Farming and Environmental Issues

The environmental argument, from the pro-organic view, holds that conventional agriculture is rapidly depleting natural resources, particularly fossil fuels and fresh water, and seriously polluting soil, water and air. Cited are the large quantities of agricultural chemicals in use, water wastage through high-volume irrigation, heavy use of petrochemicals for farm machinery and long-distance transport, high densities of various waste products from concentrated operations, and the list goes on.

While there is no argument that conventional agriculture relies on an abundance of these resources and creates a high volume of waste, agribusiness supporters argue that the negative claims are exaggerated or inaccurate. The fact that the current food industry exists and has fed the much of the world for several decades is the biggest pro-argument to date.

On the flip side, large-scale organic operations that don't follow sustainable practices would require many of the same resources as conventional operations. For example, an organic farm that made heavy use of farm machinery and indoor production facilities (requiring artificial heat and light), and shipped to far-off markets, would still be a major consumer of energy resources. Also, it is debated whether an organic farm using natural compost and manure on a large scale would cause any less damage to ground water and soil than manufactured fertilisers.

Interestingly, many organic farms rely on manure that is not organic to continue fertilisation. This technically does not violate the traditional definitions of organic produce because there are no inorganic components added to the manure, although they may be present in its composition. Studies of the effects of chemicals within manure on organic produce is limited, although studies have shown that many carcinogens are present in variable amounts in even organic foodstuffs.

Some critics, most notably Norman Borlaug, contend that adopting organic farming methods on a global scale would be more detrimental to the environment than conventional farming. Borlaug asserts that if organic farming is to feed the globe, it will require a dramatic increase in cropland area, and that achieving this goal will ultimately lead to wide-scale deforestation.

Food Contamination

Some critics point out organic food could be less safe than non-organic food, by increasing the risk of exposure to biological contaminants and food-borne diseases. In particular concerns are related to the use of manure, well known for carrying human pathogens and presence of mycotoxins from molds. One large, influential French study, evaluating organic and conventional food during 1999-2000, warned that biological toxins in certain organic products (apples, wheat) should be closely monitored.

Food contamination is usually caused by unhygienic handling and storage, including use of contaminated water, which can occur on-farm, in transit, and at the point of preparation. On the other hand, there has been no concrete evidence as of yet showing a direct link between organic farming practices and food contamination.

Food Quality

Although organic food is a topic in its own right, there are concerns related specifically to the quality of raw, fresh food. Without conclusive science either way, some organic supporters believe that the overall nutritional and health-promoting value of food is compromised by chemical-farming methods. This involves areas like micronutrients and trace elements, plant physiology, the way plants grow and the process of human nutrition.

The common sense appeal is that food grown in unnatural, sheltered, chemically assisted ways isn't as "good" for people as "naturally grown" food, as some things are different or missing. The counter-argument is that, by currently accepted standards of food science, there has been no demonstration of a functional difference between organically and conventionally produced food, and that assisted food is actually healthier and thus, more nutritious. Further, there is some concern that due to the limited methods available to organic farmers for combating quality problems while adhering to organic standards, some organic food does not generally achieve comparable safety and quality standards as "conventionally" grown products.

Soil Conservation

The practice of ploughing to prepare soil for planting is claimed to increase soil damage compared to using herbicides, like glyphosates. In fact, this argument applies primarily to large-scale, chemical-based agriculture, where huge areas are repeatedly tilled and planted with the same crops. By using artificial fertiliser rather than replacing organic material, the soil structure is progressively destroyed, and becomes increasingly susceptible to wind and water erosion.

Use of herbicides to kill weeds, instead of plowing them under, may present a short-term solution to this problem. However, repeated use of herbicides also kills microorganisms that contribute to the decomposition of plant residues that help rebuild the organic matter that holds the soil together. It also encourages the selection of the most herbicide-resistant weeds, which necessitates increased herbicide use.

Rural infrastructure

Critics condemn agribusiness practices for putting small, independent farmers out of business, destroying rural communities in the process, and causing the "art of farming" to be lost. According to these critics, small-scale organic farming encourages local economies, and provides social and employment alternatives to concentrated, energy-dependent urban living, thus improving the quality of life for everyone.

The entry of large-scale businesses into production of organic food undermines the belief that a preference by consumers for organic food will necessarily translate into a substantive change in the nature of agribusiness. This is where the distinction between organic farming, organic food, and organic certification becomes tricky. If the strong consumer trend represents simply the desire for an "organic" stamp on their food, then the trend to large-scale, global, corporate farming, certified organic or not, will continue. If consumers embrace a broader concept of "organic", which includes fresh, local food, substantial changes in the food industry would have to follow to meet this demand.

Sustainability

Although it is common to equate organic farming with sustainable agriculture, the two are not synonymous. Sustainability in agriculture is a broad concept, with considerations on many levels, such as "environmental health, economic profitability, and social and economic equity."With regard to organic farming methods, one goal of sustainability would be to approach as closely as possible a balance between what is taken out of the soil with what is returned to it, without relying on outside inputs.

An organic operation that imports the manure it uses to replace the nutrients taken out of the soil by crops, must factor in the resources required to produce and transport that manure, when calculating sustainability. Organic farming today is a small part of the agricultural landscape, with a relatively minor impact on the environment. As the size of organic farms continues to increase, a new set of large-scale considerations will eventually have to be tackled. Large organic farms that rely on machinery and automation, and purchased inputs, will have similar sustainability issues as large conventional farms do today.

Future of Organic Farming

Organic farming is at a crossroads. Despite the growth in the organic food market over the last decade, the future of the small, independent farmer, organic or otherwise, is as much in jeopardy now as it has been in recent decades. The local infrastructure to support small farmers is all but non-existent in most developed nations - the current food distribution system favors high-volume production, and large farming operations. What is commonly known as "organic farming" may change quite dramatically in the coming few years.

Organic farming is now gaining popularity and is being accepted by people all over the world. In Deborah Koons Garcia's film The Future of Food, it is stated that the American market for organically grown food amounted to $1 billion in 1994, and $13 billion in 2003. A growing consumer market is naturally one of the main factors encouraging farmers to convert to organic agricultural production. Increased consumer awareness of food safety issues and environmental concerns has contributed to the growth in organic farming over the last few years.

6

Livestock Development Strategies

Livestock make a major, although largely underestimated, contribution to rural development in developing countries. They produce food, enhance crop production and provide additional economic goods and services as well as cash income. The inclusion of livestock diversifies and increases total farm production and income, provides year-round employment and disperses risk. Sales of livestock products provide funds for purchasing crop inputs and for financing farm investments. Livestock often form the major capital reserve of farming households and, in general, enhance the economic viability and sustainability of a farming system.

Despite these positive contributions to agriculture and economic development, many formal livestock projects have failed to meet their objectives, with the result that donors are becoming more and more reluctant to support such developments. Furthermore, animal production is increasingly being viewed far more critically:

- intensive production systems, particularly in industrialised countries, are seen as a major source of pollution;
- increasing ruminant numbers in developing countries are being associated with the degradation of the rangelands and soil erosion;

— livestock development is said to favour the richer segments of society - both producers and consumers - rather than the most vulnerable;

— livestock are thought to compete directly with humans for cereal grains.

Such controversies and the inherent complexity of livestock production impose constraints that have to be addressed and which pose particular challenges not normally faced by the agricultural planner. Yet it is the very complexity of animal production systems that also offers some of the greatest opportunities for development. Livestock, because of their linkages with the overall farming system, make valuable entry points for wider agricultural development programmes. To exploit these opportunities, an integrated approach that combines both technical and institutional interventions is required.

In many countries, the difficulties associated with increasing sustainable animal production are exacerbated by limited public-sector investment and weak, ineffective support services. Programmes and projects are often poorly designed and inadequately targeted, leading to the inefficient and fragmented allocation of scarce development resources. Policies related to the livestock sector are often incoherent with ill-defined goals and with little or no assessment of their likely impact.

The lack of consistent, integrated strategies that focus limited resources on identified and attainable goals remains a major constraint to livestock development. The situation is further complicated in that livestock, especially cattle, represent wealth and status and, as a consequence, are disproportionately owned by policy-makers who have a clear vested interest - an interest that is not necessarily beneficial to livestock development in general.

There is obviously a need for an effective policy and planning framework that will optimise development resources and provide the necessary support and economic environment to allow a country's livestock resources to express its potential. This is reflected in the number of countries that have embarked on preparing strategic plans to develop their livestock subsectors, either with their own resources or, increasingly, with the assistance of governmental and non-governmental agencies.

For the agricultural planner, difficulties are related not only to the complexity of livestock production systems but also to an inability to understand how these systems function - this is primarily a problem of quantification and comprehension. One consequence is that development opportunities are often ignored, particularly the potential for using livestock as a catalyst to drive agricultural development. An understanding of the production factors and processes that affect animal production is a prerequisite for livestock development.

Production Factors

Animals themselves are the major resource, but their mobility makes them a difficult resource to quantify, especially under the extensive management systems that predominate in much of the developing world. While this may create problems for the government statistician and the tax collector, it is unique in that it allows for the exploitation of feed resources for which there are no alternative productive uses, such as extensive arid, semi-arid and cold savannahs, crop residues and agricultural fallow.

Livestock ownership is more skewed than ownership of or access to land, and, as a consequence, livestock development, especially if it concerns larger and more expensive species such as cattle, tends to produce benefits of low equity. Many non-livestock owners are often precluded from livestock development because of lack of capital or credit. Yet, in many farming systems, livestock constitute the main, if not the only, capital reserve of farming households and, importantly, one that can be readily realised. As such, livestock serve as a strategic reserve that adds stability to the overall fanning system. In this respect, a mix of species increases stability; for example, cattle represent a long-term investment whereas sheep, goats and poultry are primarily shorter-term investments and sources of petty cash.

Whereas animals can substitute for capital, purchased feeds can substitute for land, creating "landless" production systems, where land is no longer a production factor. These are not primary production systems in the strictest sense, but more industrial and almost exclusively demand-driven. Conversely, extensive pastoral systems depend exclusively on forage, the availability of which is both seasonal and highly dependent on natural factors, most notably rainfall. Such systems are essentially resource-driven and less responsive to price changes.

Many animal feeds also have alternative uses, either for human consumption or for industrial use. These competing demands are determined by price and availability of the commodity, its use and the value of the end-product. In this respect, the more intensive, demand-driven production systems are an important alternative use of these commodities, and, as such, feed-conversion efficiency becomes an important productivity indicator and management objective.

Escalating ruminant (large and small) populations, agricultural encroachment and decline in traditional authority have put an increasing strain on "open access" feed resources - particularly extensive grazing areas leading, in extreme cases, to irreversible degradation. The conflict between communal ownership of land and private ownership of livestock - the classic "tragedy of the commons" - has resulted in a disequilibrium that continues to threaten the ecological stability of many of these fragile environments.

However, an increasing body of evidence suggests that these rangeland ecosystems have adapted and are more resistant to heavy stocking than previously thought, and that opportunistic range management often is an efficient and ecologically sound resource use. For the livestock planner, these are particularly complex and sensitive issues that must be addressed. Technical options are limited, and certainly solutions cannot be found without cognisance of the wider institutional and social contexts.

Access to water, especially in the extensive rangeland systems, is a fundamental prerequisite for livestock production. Historically, water has governed both access to and Use of these resources and has provided a brake on exploitation. Major environmental issues have recently arisen over the provision of new water sources, however, particularly perennial boreholes.

Labour is the other important resource for which livestock production has a specific requirement. Typically, livestock production is more labour-intensive and less seasonal than crop production, and can utilise family labour that usually has a low opportunity cost. The distribution of labour, responsibilities and benefits tends to be favourable for women, especially with the smaller animal species, which they themselves may own. Since women are largely responsible for the day-to-day management of the household, any development programme must take their time availability into account.

Production Process

Livestock production tends to be more complex than crop production. Production cycles, although seasonally influenced, are less pronounced with livestock. Some species have very short reproductive cycles (rabbits, poultry), while those of others (large ruminants) are far longer. In more intensive production systems, seasonal influences can be offset by modifying the environment through improved nutrition and artificially controlled light and temperature.

Unlike intensive production systems, which produce a single product, many of the production systems common to the developing world produce a range of commodities. These may include a mixture of consumables and services that provide cash, subsistence and inputs (draught power and manure) into the farm enterprise. Since livestock so often have a pivotal role in the overall farm system, any constraint imposed on livestock may also restrict the system as a whole.

Animals also have an important asset function. Many smallholders, particularly in mixed-farming systems, prefer the flow products (milk, draught, manure) rather than the end-products (meat, hides and skins) since selling their animals for slaughter entails the permanent loss of flow products. Only in larger herds or flocks can meat offtake be regarded as a flow product. Flow products generate a regular cash income, unlike end-products or crop revenues, and the importance of regular cash flow, however small, is often underestimated in many development efforts, especially within the smallholder sector.

Given the perishable nature of animal products, development beyond household consumption requires marketing and processing facilities that may not be readily available. Milk, for example, not only needs a continuous outlet, but also transport, storage and processing facilities. This interdependency between animal production and the links in the marketing chain increases as systems intensify.

The close relationship between domesticated animals and humans evident throughout history continues, and many societies have strong socio-cultural values attached to their animals. In most cases, these values reflect specific economic attributes and have important implications for livestock development.

Risk is another important factor in livestock production. Extensive livestock production systems are exposed to particularly high risks as a result of natural factors such as drought and disease, whereas in mixed-farming systems livestock help to reduce the overall risks to the enterprise. As production systems intensify, the production risks from natural causes decrease and are replaced by economic risks, such as price fluctuations and taxes. Producers can mitigate risk through diversification, flexibility and increased productivity. The design of livestock development strategies must take these factors into account through risk analysis.

Consumption

On the consumption side, livestock products have characteristics that require special attention in the planning process. Meat, milk and eggs are all perishable products that need refrigeration and careful handling if they are to remain fit for human consumption. Such processes impose certain infrastructural requirements that may be beyond the resources of a given country or even a region. This is a major contributing factor to the small amount of milk products marketed in sub-Saharan Africa, for example.

Typically, livestock products also have a high elasticity of demand but a low elasticity of supply, particularly in land-based smallholder production. In many countries the demand for animal products increases rapidly with expanding urbanisation, growth in per caput income and a growing population. Because of this demand pattern it has been argued that livestock development tends to favour the higher-income sectors of society - an isolated view, yet one that has deterred potential donors - but does not adequately take account of benefits on the supply side.

All edible livestock products have high nutritive value. They are especially rich in protein, with a favourable composition of amino acids. Milk availability is one of the major contributors to the alleviation of malnutrition among the more vulnerable sections of society, most notably children. The excessive consumption of animal products, which is now recognised as a serious health hazard, is primarily a concern in affluent societies, while in developing countries consumption levels remain low. Indeed, animal fat provides a valuable contribution to the daily energy requirements in these countries.

Livestock-sector Planning

The FAO Animal Production and Health Division (AGA) is developing a methodology to provide a rational basis for livestock-sector planning that will assist governments in determining their own policies and priorities. The principles described apply equally to a particular segment, species or geographical region of the subsector-milk production, animal health or the subhumid zone, for example - and to the subsector as a whole. The chosen approach can be described as interdisciplinary, as it combines the biological, social and economic sciences, dynamic, in that it uses scenarios to anticipate future development, and systems-oriented.

In general, most livestock development strategies have similar aims, namely, to:

- — conserve the natural resource base;
- — raise productivity through better utilisation of available resources: capital (animals), land and labour;
- — expand production where there is a sufficient demand and resources can be utilised at reasonable cost to the environment;
- — optimise the allocation of development resources through rational administration and management.

A strategy for livestock development has a number of technical and socio-economic dimensions that need to be addressed if it is to have any relevance or acceptance. To make the planning process simple and transparent, the approach adopted by AGA follows the logical sequence of:

- — *Diagnosis:* An analysis of the current performance of the livestock subsector, its potentials and constraints.
- — *Prescription:* A creative phase in which programmes and policy options are formulated and designed to address identified constraints and potentials.
- — *Prognosis:* A comparative (with and without) assessment of the likely impact and implications of the prescribed programmes and policy options.

The analytical process usually proceeds from the bottom up, that is, from the lowest production unit to its effects on the national economy, and contains an ex ante evaluation of ongoing as well as previous programmes and policy interventions. The process is interactive in that the results of any comparative assessment may require modifications to either the design or further analysis.

The final result is the presentation of a series of development options along with a corresponding set of recommended actions and policies required if the options are to be implemented. Implicit in the success of any such exercise is a high degree of national "ownership" resulting from the maximum involvement of nationals in the complete sequence through such elements as rapid rural appraisals (RRAs), national consultancies and workshops and expert working groups.

With regard to livestock policy analysis and formulation, the concept of induced innovation is applied. According to this concept, the choice of technology is induced by a relative abundance of production factors, such as those introduced above, for example, in the case of livestock production.

For a number of the effects that livestock production can have, such as those on the environment and on human health, no markets exist. These effects are referred to as market failures, or externalities, which can be both positive and negative. Where these externalities exist, such as in the case of waste disposal from intensive pork and poultry production, incentives or regulations are part of policies that aim to reflect more accurately the implicit value of a non-marketed commodity, such as human health or the environment. Incentives and regulations affect factor use as they infringe on factor prices and, thus, on technological choice.

Diagnosis Phase

Livestock need to be placed within the overall context of the agricultural sector and the national economy as a whole. The first step is usually an inventory of the natural and economic resources available for livestock production: land availability and use (by agro-ecological zone); feed (crop residues, agro-industrial by-products, cereals, forages, dodders); animals (breeds, etc.); water; labour; capital; public infrastructure; and support services (marketing, input supply). Existing production methods and productivity, processing, storage and marketing all need to be assessed in relation to the available resources.

In many developing countries, traditional and commercial production systems exist alongside one another with little or no linkage between them. Industrial systems of poultry, pig and, to a lesser extent, dairy production are demand-driven and responsive to changing market conditions, due, in part, to their short productive cycles and a wide and varying range of potential feeds. Conversely, traditional ruminant production systems are inelastic in supply and sometimes even respond inversely to price increases, as was evident in the negative beef supply response observed in Latin America and in southern Africa.

A number of macroeconomic policies also have direct and indirect impacts on livestock production. Pricing policies - often through the manipulation of exchange rates - wages and interest rates determine the costs and returns of livestock production. Fiscal policies govern government expenditures on, for example, rural infrastructures and producer or consumer subsidies. Monetary policy determines the availability and cost of agricultural credit, while trade policies regulate the import and export of livestock products and animal feeds.

Institutional structures have a bearing on a wide range of support services that affect livestock production, such as marketing, quality and hygiene standards, extension and research, and agricultural education.

The balance between private and public ownership of livestock-related services is becoming a critical issue, especially as extended public services prove to be increasingly inefficient and costly to run. These macroeconomic and sectoral policies need to be analysed as to their relevance to livestock production; incentives and/or disincentives to livestock production as they affect relative abundancy of production factors; and adequacy in view of likely future developments.

In addition to these broad economic policies, countries usually pursue more specific sectoral policies that may or may not be interrelated with the overall policy framework. These policies usually refer to issues such as: land tenure; input supplies and pricing; labour supply and wages; specific livestock credits; marketing of livestock products; product processing facilities; and capacity and performance of extension, training and research institutions.

On a microlevel, policies may be analysed using econometric methods for farm household data, integrated farm household models, mathematical programming or a policy analysis matrix.

Livestock and Environment

In cooperation with eight different donors, FAO is leading a multidonor study entitled "Interactions between livestock production systems and the environment global perspectives and prospects". It addresses, as a follow-up to the United Nations Conference on Environment and Development (UNCED), the issues of livestock-environment interaction. The objective of the study is to assess the major positive and negative interactions between livestock production systems and natural resources.

The study's perspective is global, covering a range of representative livestock production systems in both developing and developed countries. Eleven production systems have been identified using a classification of agro-ecological zones and land use, which were then grouped into mixed farming systems, pastoral or extensive grazing systems and systems of landless intensive livestock production

For each of these system groups, livestock-environment interactions are rated for a number of interaction domains, including range utilisation, wildlife diversity, crop-livestock interactions, waste disposal from - production and from processing, methane emissions and domestic animal diversity. In addition, more indirect interactions, such as those related to concentrate feed demand and livestock-induced deforestation, are also appraised.

A number of the case-studies presented were developed to illustrate the livestock-environment interactions and impacts that have been observed. These case-studies will also explore approaches that can be implemented either to reduce environmental degradation or to enhance the process of building more stable production systems.

It can be concluded that product on systems vary considerably in their impact on the environment. Most beneficial effects are found in mixed farming systems, whereas landless systems are facing a range of environmental problems, mostly related to waste and resource use. Pastoral systems are mostly environmentally balanced, but this equilibrium is endangered as demographic pressure persists. Within production systems, production intensity is a major factor determining the nature of livestock-environment interactions.

Within identified farming systems, farm households can be defined according to their available (or lack of) resources, production patterns, productivity and income. Such analysis is usually derived from diagnostic surveys such as RRAs and farm management budgeting (FMB). The breakdown of a livestock subsector into various production systems is justified given the differences that exist in available resources and applied technologies, as well as in primary products and their uses, making it easier to deal with typologies.

It is at this level where most development efforts are concentrated, and in the majority of countries this usually entails a breakdown of the livestock subsector into production systems broadly based on agro-ecological zones and socio-economic patterns. Livestock production systems are basically a subset of farming systems, although their analysis must comprise the whole farm household.

Available feedstuffs may be classified by their physical properties, chemical properties or origin. Inadequate feed intake and/or unbalanced diets are usually the primary production constraint. In addition, feed is also a major cost component, especially in the more intensive production systems. Available feed resources need to be carefully assessed, therefore, with estimated feed requirements in mind.

Livestock resources consist of various species; their breeds and genetic potential, herd composition and ownership are the predominant determinants of a production system. These factors change as production systems become more an integral part of the economy. With increasing commercialisation, productivity and demand, there is a move from predominantly resource-driven systems to those primarily demand-driven.

Production systems may also be classified according to their main product(s) - meat, milk, eggs, manure and wool/fibre - as well as by services such as draught, transport and breeding. Systems may be further differentiated based on a scale ranging from "extensive" to "intensive", determined largely by their input-output relationships.

Livestock and their products fulfill different and often complementary functions. Apart from the marketed outputs, these include inputs to crop production (draught power and manure) and home consumption (subsistence), as well as their asset function and socio-cultural importance. At the farm level, therefore, livestock must be viewed in the context of the overall farm-household system, and the complex interactions involved must be understood technically as well as economically and socially.

Where livestock are an integral part of the farm household economy, flow products (draught, manure, milk), risk alleviation, capital accumulation and the utilisation of family labour with low opportunity costs are frequently the dominant production objectives. This is often in direct contrast to state objectives, which favour end-products (meat, hides, skins), and has been one reason why so many government livestock development initiatives have failed.

Strategy and Programme Formulation

A development strategy needs to anticipate and identify those forces that drive a particular livestock production system and the subsector as a whole. These can include population growth, market development, technological change and a changing resource base. An understanding of the direction and speed with which these factors may change, both at the sectoral and farm levels, is necessary so that future programmes can respond accordingly. As has been stressed, technological choice depends on the relative abundance of production factors, such as labour, land and capital.

Implicit in this process is the development of "what if" scenarios to anticipate the likely implications of any change. For demand-driven production systems, the variables usually include national and per caput incomes, population and population growth, product prices and price and income elasticities. For resource-driven production systems, trend projections that take account of anticipated changes in factor proportions, advances in technology and likely increases in productivity and resource availability are required. Any gap between supply and demand will have to be met either by expanding commercial production or by imports. The construction of scenarios is very often assisted by the use of computerised models; FAO has recently produced an updated and expanded version of this sectoral model.

The first step is to develop the "without" intervention scenario by projecting future production and consumption trends. This will indicate gaps in supply, overuse of natural resources and import requirements, among other things. The next step is the formulation of development pathways that address identified constraints, opportunities and any comparative advantages. Interventions then need to be evaluated under "with" and "without" scenarios, which involves sensitivity analysis of the critical variables as well as an assessment of social and cultural acceptability.

Historically this has primarily been a top-down approach that has failed to take account of the aspirations and production objectives of producers, resulting in the poor performance of so many livestock development programmes. It is important, therefore, to ensure the active participation of all interested parties in the formulation process.

For animal production, possible programmes and policy interventions include the following broad classifications:

— On-farm interventions, adapted to specific agro-ecological conditions and production systems. These interventions commonly aim at increasing the availability or utilisation of local feeds, control of economically important diseases (internal and external parasites) and/or improved housing and management.

— Institutional changes, including the structure and function of support services covering input supply, research, extension and training, processing and marketing and credit. Institutional programmes often complement technical interventions and aim at providing a support framework for livestock production that should be both cost-effective and congruent with overall government policies. The concept of "private" and "public" good will increasingly determine who will pay for such services.

— Genetic improvement programmes aimed at improving the livestock resource base. Options include within-breed selection of adapted indigenous breeds, substitution with exotic breeds or cross-breeding. The choice largely depends on the production system, its objectives and the resources at its disposal. Experience has shown serious misjudgement with policies aimed at importing exotic breeds with a corresponding neglect of indigenous breeds in many developing countries. Whatever breeding programme is adopted, equal attention needs to be given to the dissemination of improved genetic material. Dissemination concerns the institutional aspects as well as the choice of biotechnology, such as artificial insemination and embryo transfer.

— Animal health programmes aimed at limiting the impact of disease on animal production. Policy issues concern who will provide and pay for such services. Foremost is the need to control and protect (quarantine) the national livestock resource from major epizootic diseases, such as rinderpest, which is clearly a "public" good. Disease monitoring, veterinary investigation and legislation also fall within the public domain. On the other hand, control of diseases that cause production losses, for example, helminthiasis, is primarily a "private" good and determined on grounds of cost effectiveness. In this case, institutional policies that encourage the provision of private clinical services are required.

— Processing and marketing policies related to investment in the necessary infrastructure that enables livestock products to safely supply existing demands as well as those of the future. Such specific issues as the design of structures, equipment, training and quality, along with the question of the degree of state intervention (marketing boards, etc.) in the market, must also be addressed.

Impact Assessment

The predictive phase of the strategy design and planning process examines the various policy options and their implications in light of the government's broader development objectives. Strategy impact assessments usually address issues such as:

— economic efficiency;

— distribution and equity considerations (of both costs and benefits);

— stability (food supply, income, export earnings, etc.) and risk considerations;

— sustainability (environmental, financial and institutional);

— conformity with government objectives.

This is primarily a "bottom-up" procedure looking at the performance of the individual animal through to the production systems and finally the subsector as a whole. At the production system level, interventions can be assessed in terms of their economic, social and environmental impacts on households. It is at this level where interventions that are usually aimed at increasing productivity (mortality, fertility, carcass weights, growth rates, milk yield, etc.) have to be shown to be both technically and financially viable. The results of these production system assessments may well necessitate changes in the off-farm elements of the proposed programme, such as marketing, processing, finance, extension and research.

Programme and policy implementation will also have implications at the sectoral and country levels. And some of these implications will have an impact on programme implementation as well, where they relate to the capacity of the ancillary services and industries to support and service the proposed development. These can be manipulated through price, monetary and trade policies, institutional reform and land regulations, which are mostly under direct government control. Those under private control can only be encouraged through supportive action. The impact of a successful outcome of the proposed programmes also needs to be assessed. Increased production will have an impact on income, consumption and labour, as well as consequences for the rest of the agricultural system.

7

Feed Resources for Livestock

Rates of growth and milk production from ruminants in developing countries are generally low and often only 10% of the genetic potential of the animal. The reasons for low productivity are complex but in order of priority they appear to be the imbalanced nature of the nutrients that arise from digestion of the available forage resources when these are fed without supplements, the incidence of disease and parasitism and the often harsh climatic conditions. The use of a high yielding genotype in these situatuions is often irrelevant because the primary constraint is poor nutrition. On the other hand, resistance of particular breeds to disease and high temperatures is an important overall consideration.

Large increases in animal productivity and efficiency can be brought about by small changes in the balance of nutrients in the feed base. These increases are not restricted to laboratory studies and have been observed on a large scale at the small farm level and without alteration to the other management practices. This illustrates the large potential impact of feeding strategies in these environments. Increased production of meat and milk is only one benefit. The improved body condition of animals also has an impact on reproductive rate. The age at puberty of heifers is lowered and the intercalving interval in mature cows is decreased.Increased efficiency of forage utilisation by animals and improved reproductive rate together produce an increase in production of up to five fold without changing the basal feed resources. This is accomplished by identifying and providing critical catalytic nutrients that are deficient in the diet and therefore balancing nutrient availability with animal requirements.

In general, the supplements required are a source of fermentable N, minerals for the rumen organisms and a source of protein that is not readily degraded in the rumen but moves rapidly to the lower tract to improve the animal's amino acid supply. In many countries, these feed supplements are available. In others, there is a need to identify alternative sources and provide them in the correct amounts and in an appropriate form.In many pastoral areas of the world, the basic resource is pasture, which is generally low in total protein and fermentable N. Often a source of bypass protein is not available locally. Research and development is needed to produce supplements economically at the centres of high ruminant population densities. In the rangelands, particularly in the semi arid areas, tree forages, seeds and pods represent by far the greatest potential source of protein meals that can be used as supplements to provide soluble N, bypass protein and a source of minerals. When fed, they create a rumen environment more capable of fermentatively digesting poor quality forage.

Forage Resources

Crop and livestock production in developing countries have more than doubled in the last 30 years; however, the demand for food has increased at an even greater rate, leading to an increase in food imports of approximately

10% per year. Increase in animal production in developing countries has been brought about through increases in animal numbers. Production rates per animal have remained low or have only increased to a minor extent. The demand for food for human consumption in Third World countries is likely to increase and, since cropping land is almost fully utilised at the present time, it appears that cultivated pasture is likely to become scarcer in these countries.

The ruminant will therefore have to depend increasingly on forage biomass containing large amounts of complex carbohydrates which are not digested by intestinal enzymes and therefore require fermentative digestion. These include forage and crop residues which cannot be used by monogastric animals but can be used efficiently by ruminants.

The ability of ruminants to convert otherwise waste materials into useful products or to accomplish work will ensure they remain important livestock species in the foreseeable future.The increasing need for land for crop production indicates that ruminants will continue to use primarily crop residues, industrial byproducts and pastures from relatively infertile rangelands. The common characteristics of such feeds are low digestibility, low protein content and a low mineral components.

Animal Productivity from Feed Resources

It has been a major misconception throughout the scientific literature that low productivity of ruminants fed on forages is a result of the low energy density of the feed. This is very misleading. There is abundant evidence that low productivity in ruminants on forages results from inefficient utilisation of the feed because of deficiencies of critical nutrients in the diet. The deficient nutrients may be those critical to growth of rumen microbes which ferment or digest the feed or those nutrients required to balance the products of digestion that are absorbed to meet requirements. The evidence for these statements has been considered in detail recently by Leng. Inefficient utilisation of nutrients is accompanied by an increase in metabolic heat which has important implications for ruminant production in the tropics. The often low intakes of poor quality forage by ruminants in tropical countries may be imposed by a combination of metabolic heat stress and high environmental temperature and humidity. Correction of a nutrient imbalance by feeding a bypass protein often increases the intake of poor quality forages to between 80-100 g/kg$^{0.75}$/d, particularly in hot climates (Figure 1).

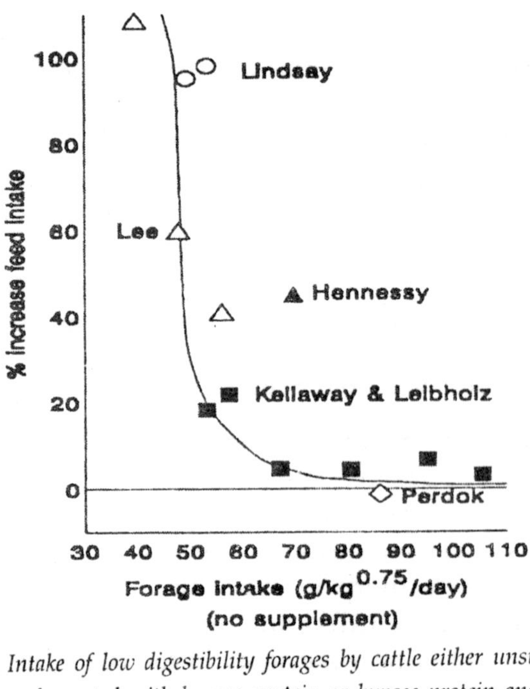

Figure 1. Intake of low digestibility forages by cattle either unsupplemented or supplemented with bypass protein or bypass protein and urea.

Cattle in the tropics may require less feed for maintenance as they do not have to combat cold stress. If they can process these nutrients that would otherwise be oxidised for maintenance of body temperature, they can be more efficient than animals in a cold climate. The energy spared, however, must be supplemented with protein to ensure an appropriate protein to energy (P/E) ratio in the nutrients absorbed for optgimum efficiency of feed utilisation. For this reason the requirements for amino acids are higher in cattle in the tropics than in animals on the same feed in cool environments.

Improving Ruminant Production

The rationale and concepts on which the following discussions will be centred have been reviewed by Preston and Leng and Leng. The basic concept is that ruminants fed low quality forages require supplementation with critically deficient nutrients to optimise productivity. The required supplements must:

- — correct nutrient deficiencies for the rumen microbes, and
- — increase the ratio of protein to energy (VFA) available from digestion so that it more closely corresponds to the animal's requirements.

Rumen Fermentative Digestion

Anerobic fermentative digestion in the rumen provides microbial cells (which supply the protein (P) to the animal) and VFAs (E) (the major source of oxidizable substrate). The efficiency of microbial growth and therefore the P/E ratio will be low when nutrient deficiencies occur in the rumen. Thus a sub-optimal level of any nutrient required for microbial growth will result in low protein to energy (P/E) ratio in the nutrients absorbed. Ensuring a nutrient non-limited microbial digestion in the rumen by complete supplementation automatically improves the P/E ratio in the nutrients available to the animal.

Feeding a protein meal in which the protein has been made insoluble or otherwise not degraded by rumen microbes is a further major supplementation strategy to adjust the P/E ratio upwards. The P/E ratio appears to be the primary factor that controls the efficiency of feed utilisation and the partitioning of nutrients into various components of production.

The ratio of microbial protein to VFA produced and the effects of supplementation to provide an optimal environment for microbial growth efficiency is less important where dietary bypass protein is completely digested in the rumen (Table 1).

Table 1. The effects on P/E ratio in the nutrients absorbed of supplementation with a bypass protein to cattle with a poor or optimised microbial milieu in the rumen. The values are calculated for a steer digesting 4 kg DM in the rumen.

Rumen environment	*Protein bypass (g prot/d)*	*Microbial cell produced (g/d)*	*Microbial protein (g/d)*	*VFA produced (MJ)*	*P/E protein (g/MJ)*
Deficient	0	830	500	41	12
Supplements	0	1630	1010	30	33
Deficient	400	830	500	41	22
Supplements	400	1680	1010	30	47

The reason for discussing these theoretical calculations at this point is to emphasise the large variations that can occur in P/E ratio (from 12 to 50) in the nutrients absorbed by ruminants fed unbalanced diets and diets balanced with supplements at the same level of feeding.

A further point is that, on a diet supplemented with a high level of bypass protein, the rumen microbes need not grow highly efficiently as long as the digestibility is optimised.

Feeding Standards

The relationship of P/E of the efficiency of feed utilisation has a very large effect on growth, milk yield and reproductive performance. The levels of production achieved when P/E is increased have been greatly superior to that predicted from present day feeding standards based on metabolisable energy of a feed. This illustrates the marked differences that result when supplements high in protein are given to cattle on diets of low ME/kg DM.

Except for temperate pasture, most forages consumed by livestock are relatively low in digestibility (55%). On most crop residues or tropical pasture, digestibility is 40-50%. The metabolisable energy in the dry matter (M/D) thus ranges from 4.8 to 7.5 MJ/kg DM. According to feeding standards, a metabolisable energy content in a feed of 7.5 will support a gain of approximately 2 g/MJ of ME intake in cattle. At 4.8MJ/kg DM cattle should be in negative energy balance. Contrast this with results of supplementary feeding trials based on balancing the nutrition of animals with urea/minerals and bypass protein, where cattle growth rates equivalent to 18 g/MJ of M/E intake have been achieved in cattle fed straw. Obviously the presently accepted feeding standards are misleading and should not be used to predict animal performance on these feeds.

The application of the basic concept of balanced nutrition, as referred to here, can improve animal growth by 2-3 fold and the efficiency of animal growth by as much as 6 fold over previous estimates. Also, although growth rates of cattle are below those on grain based diets, cattle on forage based diets in the tropics can be as efficient in converting feed to live weight gain.

The low productivity of ruminant livestock has been accepted in developing countries as an inevitable result of the poor feed base and a low feed conversion efficiency. This is no longer tenable and should give impetus to those developing animal production in Third World countries as the application of the new feeding technologies has carry-on effects of improved reproduction, increased percentage of a herd in production and an increased off take from a herd. The benefit of improved reproduction may even outstrip the direct effects on live weight gain or milk yield per se. However, the remaining challenge is to develop the necessary supplements and to get these to the animals in the various production systems.

Application of New Feeding Strategies

Even though the principles of feeding bypass protein to improve productivity have been known for many years, application has been slow and unspectacular. The application has been slowed by:

- The desire by many scientists to stand by the feeding standards that have been promulgated for twenty years and which appear to be totally inappropriate for most feeding systems.
- The controversies surrounding the principal mechanisms of action of protein supplementation which has clouded the major issues.
- The inability of research scientists to communicate with and be believed by applied technologists.
- The unavailability of protein, minerals and non-protein nitrogen in the cattle producing areas.

However, extensive application of the use of supplements high in bypass protein has occurred in India through the initiatives of The National Dairy Board of India (NDDB). For the same reasons as given above, progress was initially slow but it is now accelerating at a pace which should see most feed mills in India dedicated to the production of bypass protein supplements within the next five years. At the present time, approximately 200 MT of bypass protein feed is being fed daily to dairy animals owned by village farmers in the central parts of India. In many situations, this is coupled with the use of molasses-urea blocks compounded to provide a spectrum of nutrients for the rumen microbes. Monthly milk collection is shown in Figure 2 for a dairy co-operative in the Kedah district of India that changed from feeding a traditional concentrate to a new supplement containing 30% protein with a high bypass capacity.

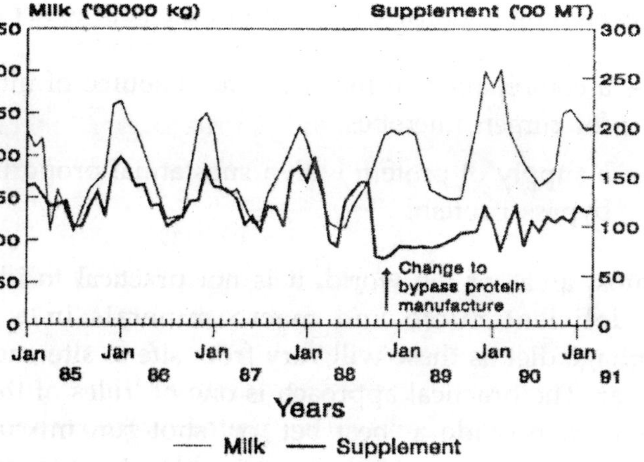

Figure 2. Milk collection records and the sale of supplements for a milk cooperative in the Kedah district of India when supplements were compounded on traditional concepts and following their replacement with a 30% C.P. by-pass protein pellet.

The collection of milk is shown for the previous five years and for the twenty-six months since conversion to the new feeding system. Whilst a number of changes have occurred in the area which could contribute to the increased milk collected, local research confirms that the responses in milk production to the new supplements are in agreement with observed increases in milk collected. The effects of the changed feeding strategies appear to be a 30-50% increase in milk production from 1 December 1988 to 1 December 1989. A further similar increase in production is apparent in 1989-1990. This increase probably represents a carry-on effect that would come from increased reproductive rate and the increased proportion of the herd in milk resulting from the improved feeding strategies.

Appropriate Strategy

To ensure maximum utilisation of local forage resources in developing countries, the appropriate supplements are:

- a combination of minerals and a source of nitrogen for rumen microbes.
- a supply of protein with a substantial proportion of bypass protein.

In most areas of the world, it is not practical to identify the deficient micro and macro minerals in a basal roughage diet as these will vary from site to site and year to year. The practical approach is one of 'rules of thumb', that is to provide a 'best bet' or 'shot-gun mixture' of minerals as economically as possible. A concentrated plant extract such as molasses provides such mixtures and can be fortified for specific areas where local knowledge points to specific deficiencies. In this respect, molasses and concentrated palm oil sludge offer useful sources of these minerals. They are also quite palatable to livestock and are useful in hiding less palatable nutrient sources in supplements. Where forages are mature and dry, a tree forage or other green forages may also provide deficient minerals.

Mineral salt mixtures are commercially available in most countries. They usually have a high content of salt and only minor quantities of the trace elements. In practice, fortified materials, such as molasses, will be superior to these mixtures as they present a greater coverage of all the minerals required and are also a valuable source of other nutrients and a small amount fermentable energy. It should also be pointed out that, where protein meals are supplemented, these often provide adequate amounts of essential minerals such as phosphorus.

Supplying the Rumen Microbes with Ammonia/urea

The other requirement is a non-protein nitrogen source for the rumen microbes, usually urea. Urea is commonly administered together with minerals and its concentration in such mixtures is controlled by safety concerns and the difficulty of incorporation, and therefore rarely exceeds 10-15% of such mixtures. However, this is usually sufficient to allow an intake of between 50 and 100g of urea by cattle from a molasses-urea block, which is sufficient to balance the rumen for ammonia on a low N roughage diet.

Results from India suggest that mineral-urea mixtures in, for example, molasses multi-nutrient blocks are best given free choice, allowing the animal some degree of selection. There are indications that the animal will learn to control the intake of urea to an optimal level. In Indonesia, the combined effects of NPN and minerals in molasses-urea blocks has a large effect on production of ruminants fed cut-and-carry green tropical pasture (Table 2).

Table 2. Effects of UMMB supplements on growth rate of Friesian Holstein steers, Ongole steers, sheep and goats fed cut/carry pasture in Indonesia.

Animals	*N (g/day)*	*Growth rate*		*% Increase in production*
		–UMMB	*+ UMMB*	
Friesian-Holstein steers	156.3	210	560	166
	171.3	400	810	102
Ongole steers	161.0	333	526	57
	204.0	478	465	-2.7
	291.0	388	822	111
	110.0	183	403	120
Sheep (local)	30.4	36	67	86
	31.5	140	316	126
Goats (does)	31.7	40	88	120
(kids)	51.7	91	105	15

Bypass Protein Supplements

Providing bypass protein to cattle owned by small farmers is often difficult and at times too expensive. There is often little information on locally available protein sources, particularly the level of protection of these protein meals from degradation in the rumen. As a rule-of-thumb, solvent extracted oilseed cakes, fish meal that has been flame dried and protein sources that have been heat treated, have some considerable protection from rumen degradation. The degree of protection is enhanced by pelleting the protein meal in the presence of free glucose or fructose, when a mild browning reaction occurs.

Identification of Protein Sources

Some countries are fortunate in having large amounts of

crop residues high in protein, most of which have a fair degree of protection brought about by processing methods. These materials are convenient for use directly by the farmer or may be processed through existing feed mills for the production of a supplement which can be fortified with minerals.

In many countries, particularly in extensive grasslands or savannas, where major constraints to production of cattle are essentially the same as those for cattle fed crop residues, protein sources may not be readily available or the sources not so obvious or easily obtainable. Most legume forages, legume seeds, edible tree leaves, seed pods and seeds that are available in these areas contain highly soluble protein which is easily fermented in the rumen. These, when used as supplements, provide a valuable source of ammonia and minerals. This increases production of cattle on a basal diet of low protein roughage but, when fed as a small proportion of the diet, they provide little bypass protein as the protein degrades rapidly in the rumen. Supplementation of cattle on green *Brachiaria decumbens* pasture with either urea/molasses or foliage of the fodder tree *Gliricidia* resulted in similar increases in production (Table 3).

Table 3. The effects of feed supplementation on livestock gain of cattle grazing on green Brachiaria decumbens pastures in the wet season with liquid molasses/urea 10% or Gliricidia foliage.

Treatment	*Rumen Ammonia (gN/l)*	*Initial Weight (kg)*	*Final Weight (kg)*	*Live weight Gain (g/day)*
No supplement	50	194	244	580
+ Gliricidia	170	204	266	717
+ Molasses/Urea	250	203	269	751

This indicates the value of a high protein tree forage as a rumen stimulator but illustrates that its effects are limited to the responses in the rumen.

Processing of Local Protein Resources

There are a number of processing techniques that will effectively and adequately treat protein sources and render them non fermentable in the rumen but allow them to retain digestibility in the intestines. In general these include chemical treatments with agents which cross link with amino acids on the protein chain, including formaldehydes and aldehydes, tannins and simple sugars such as glucose and xylose. These reactions often require the protein source to be heated in processing. Heat alone will often denature the protein and effect protection. For instance, Goering and Waldo found significant effects of the temperature of drying lucerne on the subsequent animal production from that lucerne.

In general, the higher the temperature of drying the greater the retention of nitrogen by the animal. More recently Lewis et al. demonstrated that mild heat with a small amount of xylose is very effective in protecting soya bean meal protein (Table 4).

Table 4. Effects on live weight gain of cattle supplementing a basal forage/ concentrate based diet with soyabean meal or soyabean meal treated with sulphite liquor at 200°F for 2 hours.

	Live Weight Gain (g/day)
No supplement	591
+ 7% soyabean meal	673
+ 9% soyabean meal + 10% sulphite liquor	823
+ 8% soyabean meal + 5% sulphite liquor	841

Xylose can be readily produced by acid hydrolysis of many fibrous materials including bagasse and cottonseed hulls. The effects of protecting lucerne forage protein on productivity of sheep is shown in Table 5.

Table 5. Live weight gain and wool growth of sheep fed oaten chaff supplemented with either lucerne of xylose-treated lucerne.

Treatment	*Oaten chaff intake (g/DM/day)*	*LWG (g/day)*	*FCR (g/day)*	*Wool growth (g/day)*
LX0	690	44	22	9
LX1	795	69	16	11
LX2	693	50	20	9

There is a great need to research methods for protecting protein of forages, tree leaves, seeds and pods from local resources as these are potentially available in the pasture areas of the world but must be used in the most appropriate supplementation mode.

Nutritional Potential of Fodder Trees and Shrubs

Fodder trees and shrubs represent an enormous potential source of protein for ruminants in the tropics. Until relatively recently, these feed resources have been generally ignored in feeding systems for ruminants, mainly because of inadequate knowledge on various aspects of their potential use, as well as initiatives associated with the development of more innovative systems of feeding.

Throughout the tropics, and especially in the humid regions, there exist a variety of feed resources. These include a variety of forages and abundant supplies of crop residues, agro-industrial by-products and also non-conventional feed resources. Among these, approximately 50 to 60% of the feeds produced are dry bulky roughages, mainly cereal straws and stovers. In Asia, the volume of production is increasing at about 3% annually. However, not all these feeds are put to maximum and efficient use. Inefficient utilisation is identified with low levels of animal production in which the contribution from ruminants compared to non-ruminants is especially low. The use of fodder trees and shrubs has been secondary to these efforts, despite their potential value in prevailing small farm systems. These alternative feeds merit increased research and development in the future.

It is appropriate to keep in view several fundamental aspects of nutrition which are associated with the use of tree fodders and shrubs. These are voluntary feed intake (VFI), which determines the amount of dry matter intake (DMI), and digestibility.

Dry Matter Intake

Maximum DMI is very important factor in ensuring the release of adequate nutrients for maintenance and production. With temperate grasses, voluntary food intake (VFI) directly affects their digestibility, but the relationship is less definitive fór tropical species because of the different lengths of time required to digest tropical feeds. VFI has been shown to decrease with decreasing digestibility of dry matter within species for *Chloris guyana*, *Panicum* spp. and also for legumes. Minson also reported that tropical grasses decrease in dry matter digestibility at a daily rate of 0.1 to 0.2 digestibility units.

In terms of VFI, considerable variation exists between and within tropical grasses. Some of the variations are due to differences in digestibility but other unrelated factors such as those recorded for *Panicum* varieties may be involved. Rate of decrease in digestibility of younger tropical herbages is as high as in temperate species. The decline in digestibility with age of tropical grasses was more rapid than with tropical legumes which retained relatively high digestibilities at maturity. Differences in *in vitro* digestibility have been reported between genotypes of *Digitaria*. Selection for high *in vitro* digestibility was successful in producing a high *in vitro* digestibility of dry matter and superior VFI of *Cenchrus ciliaris*.

In the humid tropics, DMI is limited by the water content of, or the free water on, the ingested herbage. In the West Indies for instance, the dry matter content of herbage during the wet season was very low in Pangola grass (*Digitaria decumbens* Stent) with a dry matter content of 23.4%, compared to 39.3% in the dry season, such that the herbage contributes a high proportion of the total water consumed. Similar observations have also been made in Thailand. Inadequate dietary energy arises from reduced DMI and is likely to occur when the DM content falls below 25%.

Digestibility

Associated with VFI of the forage is digestibility. Digestibility of a feedstuff is affected by stage of maturity of the crop, botanical composition, DMI and dietary supplements, processing, and chemical treatment. In general, the more digestible a feedstuff is, the more it is eaten by the ruminants. Such is the case that high digestibility increases DMI. Increasing digestibility means that high proportion of the food is absorbed and digestion is more complete, with the volatile fatty acids (VFA) showing a lower proportion of acetic acid, the energetically lowest and least useful VFA, and with the presence of higher proportions of the more useful acids such as propionic and butyric acids.

Level of Dietary Protein

VFI is influenced to a very large extent by the dietary crude protein content. The protein content of tropical forages is generally low. The protein content falls rapidly with growth and reaches a low level before flowering. During the dry season, the crude protein levels fall to very low critical levels, even below 7% in the dry matter. The level of protein in the diet affects voluntary intake of food and low protein diets are not readily eaten by ruminants. In sheep, a 7% crude protein begins to limit intake. Leibholz and Kellaway have estimated that the minimum required crude protein of a poor quality diet with a digestibility of organic matter of 50% would be between 6.1 to 7.4%.

Quantity of Net Energy

The quantity of net energy (NE) available for production is controlled by three related factors: the quantity of feed eaten (I), the proportion of each unit of feed that is digested (D), and the efficiency of utilisation of the products of digestion (E). This is represented by the equation:

$$NE = I \times D \times E$$

The quantity of NE available and its utilisation is dependent on the type of feed, nutrient balance, extent of VFI, animal species, and function.

Nutritional advantages of Fodder Trees and Shrubs

There is considerable diversity in the forage supplements that are of value to ruminants. Table 6 brings together the more commonly used shrubs and tree fodders in different parts of the developing countries. The list is not exhaustive, but is meant to focus on those feeds that are emerging as important in feeding systems for various ruminants. Of the 12 more important feed sources listed, *Acacia* spp., *Ficus* spp., cassava (*Manihot esculenta* Crantz), *Erythrina* spp., *Gliricidia* spp., *Leucaena leucocephala*,

Prosopis cineraria and *Sesbania* spp. are widely used in the tropics.

The value of these forages is associated with a number of advantages. With *Leucaena* for example, it provides a valuable source of protein, energy and sulphur for the rumen bacteria. It is also valued in multipurpose use in fence lines and as a fuel. With specific reference to their value in animal feeding, the advantages include *inter alia*: availability on the farm; accessibility; provision of variety in the diet; source of dietary nitrogen, energy, minerals and vitamins; laxative influence on the alimentary system; reduction in the requirements for purchased concentrates; and reduced cost of feeding.

Table 6. Important tree fodders and shrubs for feeding ruminants.

	Common name	*Botanical name*
1.	Acacia	*Acacia catechu*
	Acacia nilotica	*Acacia siberiana*
2.	Cassava	*Manihot esculenta Crantz*
3.	Calliandra	*Calliandra calothyrsus*
4.	Erythrina	*Erythrina variegata*
5.	Ficus	*Ficus exasperata*
	- Banyan	*Ficus bengalnensis*
-	Peepul	*Ficus religiosa*
6.	Gliricidia	*Gliricidia sepium*
		Gliricidia maculata
7.	Jackfruit	*Artocarpus heterophyllus*
8.	Leucaena	*Leucaena leucocephala*
9.	Pigeon pea	*Cajanus cajan*
10.	Prosopis	*Prosopis cineraria*
11.	Sesbania	*Sesbania grandiflora*
		Sesbania sesban
12.	Tamarind	*Tamarindus indica*

Nutritional characteristics

Table 7 presents some of the nutritional characteristics of principal tree fodders and shrubs. The data has been assembled from various sources, but in particular Devendra, N.R.C., Kearl and Devendra. In the absence of data on cell wall contents, the average crude fibre level is indicated.

The table indicates that the crude protein (CP) contents of many of these feeds is high and in the range 22.2 - 25.8%. *Ficus, Acacia, Gliricidia,* jackfruit and *Prosopis* had crude protein contents in the range 14.0 -15.1%. The metabolisable energy (ME) content, with the exception of acacia was in the range 11.2 - 14.4 MJ/kg. It is of interest to note that the Ca content of many of these feeds is also relatively high. These include, in particular, cassava, *Calliandra, Prosopis,* tamarind and

Gliricidia, in which the range was 1.57 - 2.81%.

Table 7. Nutritional characteristics of principal tree fodders and shrubs.

Feed source	DM (%)	% DM basis				ME (MJ/kg)	Ca (%)	P (%)
		CP	CF	EE	Ash			
Acacia	29.0	15.1	22.6	8.9	8.2	8.4	1.21	0.06
Cassava	21.1	24.2	15.6	4.0	6.6	14.4	2.62	0.22
Calliandra	26.4	24.0	21.7	2.4	8.0	12.6	1.6	0.2
Erythrina	32.0	25.8	17.4	5.8	6.7	14.3	-	-
Ficus	17.0	14.0	22.4	4.5	5.8	12.0	1.31	0.17
Gliricidia	25.0	14.7	19.9	5.4	4.7	12.84	1.58	0.29
Jackfruit	36.6	14.0	22.1	3.8	11.5	14.2	1.46	0.15
Leucaena	30.0	22.2	19.6	6.9	4.4	12.1	0.27	0.12
Pigeon pea	25.2	22.8	20.1	5.6	5.8	13.4	0.37	0.17
Prosopis	23.4	14.0	17.8	1.9	6.8	11.2	2.73	0.15
Sesbania	18.0	22.6	18.4	2.1	9.3	13.6	1.48	0.34
Tamarind	28.0	14.0	21.0	4.6	8.6	14.4	2.81	0.20

Beneficial Effects

Tables 8 demonstrates the benefits in buffaloes and goats concerning feeding systems in which cereal straws were used. It enabled the following main conclusions:

i. The use of forage supplements consistently increased live weight gain or milk production.
ii. Of the forage supplements that have been used, the leguminous types have been especially advantages.
iii. The beneficial responses (meat or milk production) was associated with a reduced cost of feeding.
iv. With large ruminants, research on the utilisation of forage supplements for draught is very sparse.

Table 8. Results of economic benefits of forage supplements in diets for buffaloes

Feeding regime	*Forage supplement*	*Location*	*Significant response*	*Result*
Wheat straw + conc.	Lucerne + berseem	India	Milk	Reduced cost/kg SCM milk
Rice straw	*Gliricidia* or *Leucaena*	Sri lanka	Milk	Increased margin over costs
Rice straw	Berseem	India	Milk	Reduced cost/kg milk
Rice straw	*Gliricidia*	Sri Lanka	Milk	Increased milk production

In Sri Lanka, a supplement of 1.5kg of rice bran with 3.0kg *Gliricidi· maculata* per head to Surti buffaloes on urea treated rice straw in two systems of production showed that supplementation significantly increased (P>0.05) the difference in milk and butterfat production. There was no effect on live weight gain of cow and calf. Earlier work, also in Sri Lanka, showed that feeding

1600g *Gliricidia sepium* forage (DM/cow/day) increased milk and milk fat yields. With coconut cake supplementation, both milk yield and milk fat yields were significantly increased (P>0.05), whereas *Gliricidia* and *Leucaena* leaves did not affect either component.

Supplementation of *Chloris gayana* hay with 200 to 300g of dried *Leucaena* leaves given to goats in Tanzania significantly increased organic matter intake, crude protein intake and daily growth rate compared to the unsupplemented control group. In the digestibility study, *L. Leucocephala* supplementation increased total dry matter intake from 42.3 kg W^0 in the control group to 77.9 kg $W^{0.75}$ in hay with leucaena fed *ad libitum.*

Current evidence of the value of supplementary forages and especially the leguminous ones suggests the following pointers for practical application:

— Optimum dietary level on DM basis: 30 to 50%

— As per cent of live weight: 0.9 to 1.5%

Potential Technologies

The extension of the value of tree fodders and shrubs for ruminants needs to be identified with practical technologies that can ensure demonstration of the beneficial use and also relevance at the farm level. The following practical technologies are worthy of attention and are potentially important:

(i) three-strata forage system;

(ii) integrated tree cropping systems;

(iii) agro-forestry systems;

(iv) food-feed intercropping;

(v) relay cropping;

(vi) alley cropping; and

(vii) grazing and stall feeding systems.

Three strata forage system

In dryland farming areas, a major constraint to higher productivity from ruminants is the unavailability of good quality feeds especially during the dry season and periods of drought. The development of feeding systems that can increase the supply of good quality forages and dietary nutrients for the animals is therefore especially important to improve the prevailing low level of animal performance.

This is exemplified by the situation in the island of Bali which has approximately three million people and three rainfall zones. Twenty five *per cent.* of the land area is semi-arid with a rainfall of 900–1500 mm/year. Farmers constitute about 70% of the total population and most of them practice mixed crop-animal farming. Among ruminants, Bali cattle are particularly important and in the dry parts of the island, income from livestock accounts for about 39–43% of total farm income. Farmers generally own 2–3 head of cattle which are used for draught and beef production.

The need for increasing the feed resource base as well as develop sustainable systems of production led to the successful demonstration of the three strata forage system. The system involves grasses and ground legume (first stratum), shrub legumes (second stratum) and fodder trees (third stratum).

Considerable progress has been made in the implementation of this concept and research over the last 6 years has demonstrated considerable benefits. For example, the inclusion of *Stylosanthes, Centrosema, Acacia, Gliricidia* and *Leucaena* increased the supply of forages and enabled higher stocking rates and live weights to be achieved: 3.2 animal units (375 kg)/ha/year in the TSFS compared to 2.1 animal units (122 kg)/ha/year. Additionally, the system promoted increased firewood production as well as reduced soil erosion.

Integrated tree cropping systems

Integrated tree cropping systems with coconuts, oil palm and rubber trees, involving ruminants, is a potentially important production system especially in the humid tropics. The benefits of the system include: increased fertility of the land via the return of dung and urine; control of waste herbage growth; reduced use of herbicides; presence of shade reduces heat stress; easier management of the crop; and distinct possibilities of increases in crop yields, consistent with greater economic including sale of animals and their products.

In these situations, tree crops have a very valuable role in terms of multi-purpose use: supply of feed, fenceline, fuel and soil fertility.

Silvi-pastoral systems: Agro-forestry systems that combine the use of multi-purpose trees and integration of animals is an avenue that has not adequately investigated. The system has considerable importance, especially in upland areas and elsewhere where the role of fodder trees becomes especially important. The wider use of agro-forestry systems has complimentary advantages of forage production, supply of fuelwood, improvement of soil fertility and permanent soil cover and promotion of economic and sustainable land use.

Food-feed intercropping: The concept of food-feed intercropping in both lowland and upland small farm systems has two principle advantages: firstly, the system aims to provide sustainability, involving the complimentary role of crops and animals; secondly, and with respect to animals, the use of appropriate forage crops provides fodders and crop residues which are valuable for feeding both ruminants and non-ruminants, as has been demonstrated by work in the Philippines.

The criteria for the choice of the crops to be used in the system include *inter alia*: the type of animals reared, potential forage or crop residue biomass yield, promotion of soil fertility, drought tolerance and extent of the dry season, shade tolerance in upland areas, ease of eradication and resource requirements. The strategy is to integrate, within the rice cropping pattern, other feed producing crops and also forage crops without, reducing the area of the land used.

Relay cropping: An important means to further increase the supply of feeds for farm animals is to plant a second crop into the first before harvest. Examples of relay cropping are the introduction of legumes (e.g., groundnut and pigeon pea) into the main crop (e.g., rice or wheat). The strategy extends the supply of feeds throughout the year.

Alley cropping: Alley cropping provides an alternative opportunity to integrate crops with animals as well as enhance sustainability. The development of the system has the following advantages: the maintenance and promotion of soil fertility; increasing the supply of animal feeds and fuelwood; and contributing to the development of all-year-round feeding systems.

Grazing and stall feeding systems: Increased and more intensive utilisation of tree forages and shrubs is largely dependent on the type of feeding systems to be practised. Where land is not limiting for intensive pasture production, either grazing and or stall feeding can be adopted. A comparative study of both systems to examine potential milk production in Sahiwal x Friesian cows on *Leucaena leucocephala-Brachiaria decumbens* mixed pasture indicated that rotational grazing was better than stall feeding with mean responses of 8577 and 9180 kg/ha/lactation. Supplementation of concentrates at 4 and 6 kg/cow/day further increased milk production to 13323 and 17070 kg/ha/lactation respectively. The net profit per cow with or without supplementation was lower for rotational grazing on account of higher labour cost for the stall feeding system. The choice of one or the other system is dependent to a very large extent on the cost of the inputs.

Deleterious Principles

An area which is emerging to be extremely important in the utilisation of shrubs and tree fodders is the problem of deleterious principles, since they constitute problem in the diet of animals. A variety of these principles are found.

Tannins exist in the condensed or hydrolysable forms at varying levels. The effects of tannins in ruminant feeding is not entirely clear, with possible harmful and beneficial effects, but evidence is increasing that they can have some benefits. High concentrations of tannins can lower voluntary feed intake. More recently, Feng Yu and Leng have shown that low concentrations of condensed tannins from *Acacia floribunda* provided protection of proteins in lupins and stimulated growth and wool production.

There appears to be an inverse relationship between the proteinprecipitating capacity of tannins in tree leaves and their palatability, voluntary intake and digestibilities of crude protein and dry matter in grazing and also browsing animals. The authors have also reported that sulphur and iron become limiting to animals consuming tannin-rich leaves and prolonged consumption of these also induces toxicity. Several of these feeds have deleterious principles and depending on the levels fed, some have side effects on animals. In general, however, these do not present a major barrier to their utilisation by ruminants when used as supplements.

Under these circumstances, the presence of tannins is an advantage as they protect the proteins from excessive microbial degradation and make these more available posterior to the rumen. Some experiments have shown that supplementation of a basal diet of tropical grass with dried or fresh *Leucaena* resulted in significant quantities of protein leaving the rumen of sheep.

Farmers currently overcome and reduce problems of effects on animals by feeding mixtures of the feeds with or without sun drying. This process not only extends choice of feeds available but also dilutes and reduces problems of palatability and side effects. More information is required on optimum dietary levels of forage supplements either fed alone or in mixtures appropriate to individual ruminants, methods to reduce the incidence and development of suitable mixtures of these in economic feeding systems for individual ruminants. It is possible that between species differences also exist in the response to tannin content in these leaves and these need to be determined.

Fodder Plants as Protein Sources in Monogastric Nutrition

Over the past two or three decades, there has been a significant upsurge of interest in the use of fodder trees and shrubs as sources of protein and other nutrients for non-ruminant animals. Although considerable attention has been given to the use of leaf meals derived from these trees and shrubs, there is a growing realisation that the seeds of these species may serve as protein sources in their own right and limited work has been published in this area.

Much attention has been focused on leaf meals from *Leucaena leucocephala* and *Manihot esculenta* (cassava) but, more recently, interest has turned to leaf meals derived from other sources such as *Gliricidia sepium, Robinia pseudoacacia* (black locust or false acacia), *Cajanus cajan* (pigeon pea), *Sesbania sesban, Prosopis* spp. (mesquite) and *Albizia* spp. It is clear that the leguminous species predominate in this group due to their relatively high crude protein (CP) contents and to the ability of some species to thrive in adverse soil and climatic conditions. While the potential exists for the wider utilisation of the seeds of fodder trees in non-ruminant diets the limited data in the literature relate to the grain of just one legume: *Cajanus cajan*.

Feeding trials with leaf meals

The relatively high crude protein (CP) content of leaf meals in comparison with that of cereals is well recognised. However, the fibre component is also a major fraction of the dry matter (DM). In many instances the fibre content of leaf meals may equal or exceed CP concentrations. Consequently, digestibility of the CP fraction of many leaf meals is low which tends to depress overall CP digestibility when leaf meals constitute a significant proportion of the diet.

Although lysine concentrations in leaf meals are considerably higher than those of cereal grains and certain by-products such as coconut oil meal, they are somewhat inferior to those of soyabean meal and fish meal. It follows that leaf meals cannot be expected to fully replace high quality ingredients in diets for monogastric animals. Deficiencies of the sulphur containing amino acids add a further dimension to the nutritional limitations of leaf meals. For example, biological values (BV) for cassava leaf meal range from 0.49 to 0.57 which may be enhanced to 0.80 on supplementation with methionine. Although this leaf meal confers a superior dietary amino acid profile relative to by-products such as coconut oil meal, this difference is not reflected in BV determinations with pigs.

Further insight into the nutritional value of leaf meals is provided by a consideration of metabolisable energy (ME) content. The limited data point to extremely low ME values for *Leucaena* and *Robinia* as determined with poultry, but the ME content of cassava leaf meal appears to be appreciably higher, not only for poultry but also for pigs.

Leucaena: Consistent with the low digestibility of CP and the inadequate ME content, graded additions of *Leucaena* leaf meal induce dose-related depressions in growth of chicks even when maize oil is used to compensate for the low ME value of the leaf meal. The adverse effects of *Leucaena* on the growth of young chicks is now well established. Moreover, with older broiler chicks, D'Mello *et al.* showed that diets containing 100g leaf meal/kg significantly reduced growth without affecting DM intakes. On the other hand pigs given diets containing *Leucaena* leaf meal at 100 g/kg grew significantly faster than control animals but live weight gain declined progressively with higher inclusions. It should be noted that in this study *Leucaena* was substituted for a commercial ration containing 180g CP/kg diet and, consequently, all *Leucaena* diets had higher CP concentrations than the control diet. Furthermore, all diets were supplemented with ferrous sulphate to counteract possible toxicity arising from the non-protein amino acid, mimosine, present in the leaf meal. The addition of ferrous sulphate to the mimosine-free control diet was unnecessary and may have depressed growth of pigs on this diet.

Despite the favourable digestibility of the CP fraction of *Leucaena* leaf meal for rabbits, graded additions of this legume to a control diet precipitated progressive depressions in growth such that, at the 400g/kg inclusion, weight gain was less than 40% of control values .

Concomitant with these growth responses is a consistent and striking increase in feed:gain ratios for the three classes of animals when dietary inclusion rates of *Leucaena* leaf meal exceed 200g/kg (Figure 3).

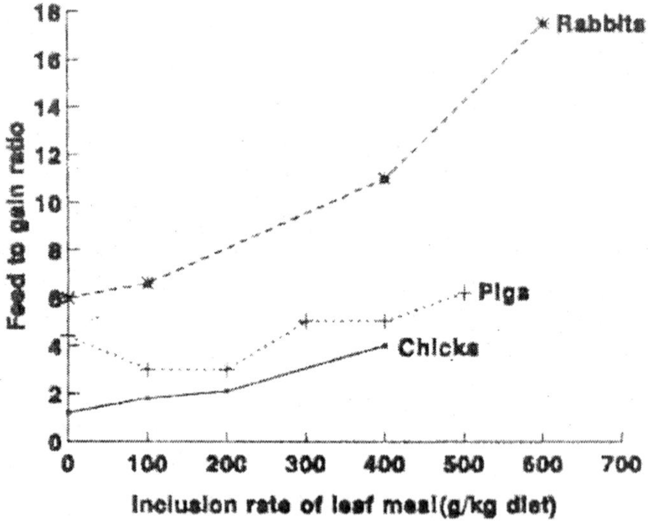

Figure 3: Effect of Leucaena leaf meal on feed to gain ratios of growing chicks, pigs and rabbits.

The data for chicks relate to two separate studies by D'Mello and Acamovic and D'Mello *et al.* in which oil additions to diets were held at a constant level irrespective of inclusion rates of *Leucaena*. Although the studies of Malynicz show significant improvements in feed conversion ratios on inclusion of *Leucaena* at rates of up to 200 g/kg diet, more recent studies by Chen *et al.* indicate no differences in these ratios between the control diet and diets containing up to 160g leaf meal/kg. The decrease in feed efficiency of rabbits given high *Leucaena* diets is all the more remarkable given the herbivorous nature of this species. It is apparent that coprophagy confers no special advantage as regards the ability of rabbits to utilise *Leucaena*-based diets.

Egg production in hens is also adversely affected by dietary inclusion of *Leucaena* leaf meal. For example Vohra et al. showed that while inclusion rates of 50 and 100g/kg diet elicited satisfactory laying performance, the diet containing 200g/kg precipitated severe reductions in food intake and egg production. These deleterious effects occurred despite adjustments to the ME content of the diet by addition of oil. It should be noted that the duration of this study was only five weeks. Earlier studies over eight weeks employing pair-feeding techniques indicated that the depression in egg production was specifically attributable to the intake of *Leucaena* leaf meal.

Cassava: Although cassava leaf meal has been accorded with a favourable ME value, graded inclusion of this leaf meal as a replacement for soyabean meal and maize in practical-type diets elicits progressive depressions in growth rates of both growing chickens and pigs. In addition, cassava leaf meal induces greater reductions in growth rate of chickens than dehydrated alfalfa meal at all inclusion levels from 50 to 200 g/kg diet. However, if substitutions of cassava leaf meal are accomplished at the expense of poor quality ingredients such as coconut oil meal in a control diet, then dietary inclusion rates of cassava leaf meal of up to 150g/kg may be employed without reducing growth of broiler chicks. Similarly, Ravindran *et al.*, showed that performance of growing pigs was unaffected by cassava leaf meal concentrations of up to 267g/kg diet when the leaf meal was substituted for coconut oil meal. At the 400g/kg dietary inclusion, cassava leaf meal significantly reduced average daily gains of pigs. A different picture emerges, however, if this leaf meal is used as a replacement for good quality protein sources. Thus, Ravindran substituted cassava leaf meal at rates of 100, 200 and 300g/kg diet for maize and soyabean meal and observed significant reductions in live weight gains of pigs with each level of the leaf meal. The effects of cassava leaf meal as a replacement for soyabean meal and maize are most clearly seen in the feed: gain ratios depicted in Figure 4.

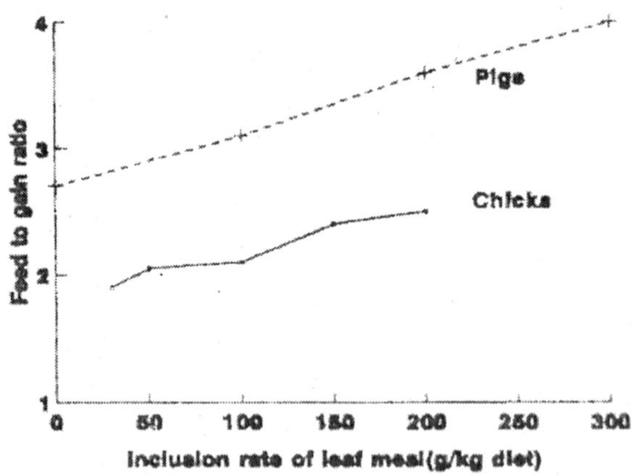

Figure 4. Effect of cassava leaf meal on feed to gain ratios of growing chicks and pigs

Other leaf meals: Limited work with leaf meals from *Robinia, Sesbania* and *Gliricidia* suggests poor replacement values for conventional ingredients in diets for poultry. In the case of *Sesbania*, there are signs of acute toxicity with inclusion rates as low as 100g/kg diet, with 100% mortality in chickens fed *Sesbania* at 300g/kg diet. Pigeon pea leaf meal may be more acceptable as a feed component for laying hens. However, feed efficiency declines on inclusion of this leaf meal and body weight losses may also occur.

Potential of grain from fodder trees

On the basis of chemical composition, the grain of fodder trees should offer greater potential as protein and energy sources for non-ruminants than do their corresponding leaf meals. Thus CP concentrations have been reported to be over 30% higher in the seed than in the leaf of *Leucaena leucocephala* and, although CP concentrations in pigeon pea are similar in the seed relative to the leaf, fibre content of the former is considerably lower leading to favourable ME values for poultry.

The feeding value of pigeon pea seed is reasonably well documented. In one investigation, live weight gains of broilers were unaffected by dietary inclusion of raw pigeon pea seed even at levels of 500g/kg. Pigeon pea seed was substituted for maize and soyabean meal in the control diet. Feed conversion efficiency, however, deteriorated with diets containing 400 and 500g seed/kg. It should be noted that ME values for the seed were somewhat low at around 8 MJ/kg DM and consequently graded levels of oil were added with increasing levels of pigeon pea seed. On the other hand, the results of two trials with laying hens indicate that raw pigeon pea seed induces adverse effects on egg production and feed efficiency when inclusion rates of 300–400 g/kg diet are used. Body weight losses occurred consistently with diets containing 400g seed/kg, a response similar to that observed with lower dietary concentrations of the leaf meal of pigeon pea. Deleterious effects also occur in pigs fed raw pigeon pea seed at 300g/kg diet.

The nutritive value of seeds from other fodder trees and shrubs has not been examined on any systematic basis but potential exists for exploitation of a variety of seeds, particularly those from leguminous browse species.

Upgrading of lead meals and grain

It will be apparent from the foregoing account that the observed nutritive value of leaves and seeds of fodder trees frequently falls considerably short of that expected from their chemical composition. The constraints to enhanced utilisation of these commodities reside chiefly in factors such as fibre content, the presence of anti-nutritive compounds and deficiencies of certain essential amino acids.

The lower fibre contents of seeds relative to leaf meals from fodder trees inevitably means superior quality of the former. Thus the seed of pigeon pea has an ME value for poultry and pigs, which is considerably in excess of the ME content of any of the leaf meals examined thus far.

The leaves and seeds of fodder trees and shrubs are known to contain a wide array of compounds which are capable of reducing performance of animals. A complex combination of different groups of compounds such as toxic amino acids, tannins, cyanogenic glycosides and proteins may occur in a single tree species. The presence of such a disparate range of deleterious compounds has provided the stimulus for much work on methods of detoxification and upgrading of both leaves and seeds. The production of leaf meals for non-ruminant feeding inevitably involves at least some measure of processing. Sun-drying is the method of choice in the tropics and this treatment alone can be an effective method for the removal of anti-nutritive substances in some leaf meals. Thus, sun-drying of cassava leaves may reduce HCN concentrations by as much as 90%.

Thermal processing is an effective method of upgrading plant products through the inactivation of most of the heat labile anti-nutritive substances. In this respect, Visitpanich *et al.* were able to reduce trypsin inhibitor activity by heating the grain of pigeon pea to 124°C for 15 minutes. This treatment eliminated all adverse effects on performance of pigs fed 300g seed/kg diet. Heat treatment is a less effective procedure for the upgrading of leaf meals derived from *Leucaena, Robinia* and *Sesbania* and, in the case of cassava leaves, boiling may reduce the availability of amino acids.

Another strategy for upgrading leaf meals involves the use of additives. Ferric sulphate and polyethylene glycol (PEG) have been used to complex with mimosine and tannins respectively, with marked improvements on the growth of chicks fed *Leucaena*-based diets. Methionine supplementation appears to offer the most viable procedure for upgrading cassava leaf meal for poultry and pigs, in combination with energy-rich additives such as coconut oil. The response to methionine is presumably a reflection not only of the low content and availability of methionine but also of the need for this amino acid in detoxification of residual HCN.

Feeding Behaviour of Ruminants

Selection and intake of diet depends not only on the available plant resources but also on the feeding behaviour of the animals. Better understanding of feeding behaviour allows the development of management strategies aimed at maximising the use of ecosystems for increased animal production. Among domestic ruminants, camels are classified as browsers, goats as intermediate selective feeders with preference for browse, sheep as nonselective intermediate feeders with preference for grasses and buffalos, cattle and donkeys as grazers. These dietary preferences are however influenced by the constantly changing environment and human intervention in livestock management.

Most behaviour studies have been carried out by visual observation of grazing activity but also by examination of faeces, rumen contents and intake samples collected through oesophageal fistulae and evaluation of the forage offered minus the quantity left over. These studies have shown that feeding behaviour is controlled by factors of animal and plant interaction. The animal factors including species, physiological status, previous experience with the vegetation and habitat exploration, while the plant variables relate, among other things, to availability and the physical and qualitative characteristics of plant biomass. In grazing situations, the relative importance of all these factors can be modified by environmental variables such as human interventions which determine strategies of land use and stock management.

Some results of the behaviour studies carried out with cattle, sheep and goats by one of the authors in the arid and semi-arid programme of International Livestock Centre for Africa will be used here to describe the complex interaction between animal, forage availability and management decisions.

The study site in the sahelian zone of Mali has an average annual rainfall of 580mm occurring during the July-September period. The notable features are: (i) an irrigated land of 45,000 ha devoted to rice cultivation where a few trees and shrubs are found along irrigation ditches as well as around villages; and (ii) a dryland with villages and millet fields. The behaviour studies concerned:

1. A herd of cattle under a semi-sedentary management system which associates the exploitation of both the dry and irrigated lands. Over the year the animals stayed successively on the rangeland (July October), millet fields (November-December) and rice fields (January - June).
2. A herd of cattle subjected to sedentary management on the dryland. The animals therefore had a fixed settlement all the year round and they exploited the rangeland from July to September, and the rangeland plus millet fields from October to June. The feeding behaviour of these animals is presented in Figure 5.
3. A herd composed of sheep and goats under a sedentary management system on the dryland. The feeding of these animals was based on range resources and millet residues as for the sedentary cattle above.
4. A herd of sheep and goats managed under sedentary system on the irrigated lands. During the rainy season, the feed resources were limited to grasses and shrubs along irrigation ditches and, in the dry season, they were extended to the rice straw and regrowths of grasses on harvested rice fields.

The preference of the cattle for grass rather than for browse is obvious in Figures 5 and 6, particularly during the wet season when the cattle spent more than 95% of their grazing time on grass despite the simultaneous availability of abundant browse with the grass. Browsing started to increase at the beginning of the dry season (October) but this intake was not sustained by the cattle under semi-sedentary management because of their shift to millet, then to rice fields where there were relatively few trees and shrubs. Conversely, with the cattle under the sedentary system, browsing in the dry season increased steadily, with a peak at the close of the season (June) because of the flush of a great number of trees.

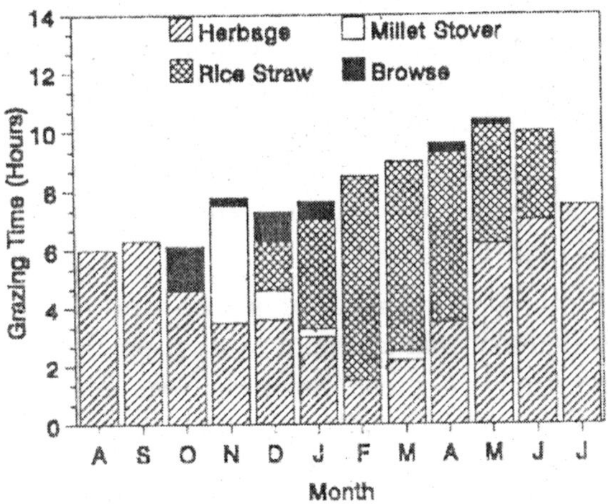

Figure 5. Semi-sedentary cattle.

The increase in the grazing time during the dry season in Figure 5 and the reverse trend in Figure 6 are due to the influence of herd management on grazing activity. In the semi-sedentary system, the herdsman always settled on an open pasture near a watering point convenient to both his family and cattle, while in the sedentary system the permanent kraal and watering point had an eccentric position relative to pastures.

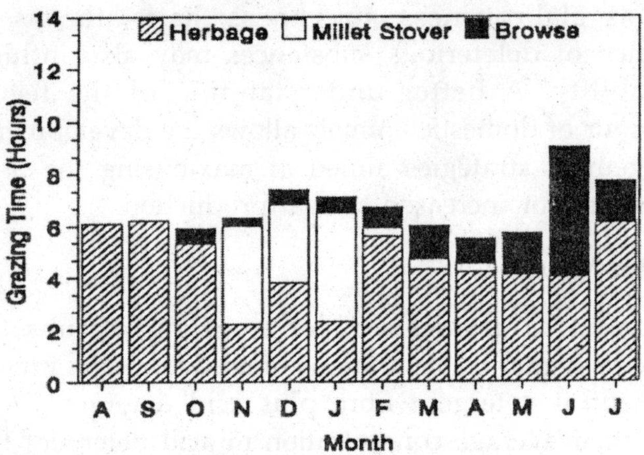

Figure 6. Sedentary cattle in dry areas.

In consequence, the semi-sedentary cattle spent less time walking to and from pastures than did the sedentary animals. In addition, they were able to practise night grazing because of the closeness of the pasture lands while in the sedentary system, long walks and lack of night grazing had a depressive effect on grazing activity.

Among domestic ruminants, the most effective browsers are camels. In the arid zones of Kenya, Lusigiet al. have found that more than 50% of the tree species and 55% of the shrub species were utilised by camels, while only 14 to 19% of the perennial grasses were eaten. There is a great overlap between the diet of camels and goats but little competition in browse utilisation occurred between the two species owing to the superior reach of the camels.

Preference for specific plant species and seasonal changes in palatability of trees and shrubs are also some features of the feeding behaviour of domestic animals. In addition, some intrinsic characteristics of trees and shrubs such as glabrousness, toughness, hairs, thorns and presence of deleterious substances may also influence palatability. A better understanding of the feeding behaviour of domestic animals allows the development of management strategies aimed at maximising the use of ecosystems for increased animal production.

Nutritive Value

Compared to grasses, fodder trees and shrubs have relatively higher concentrations of crude protein, minerals and neutral detergent fibre plus acid detergent lignin, while their average concentration in acid detergent fibre, as well as their average dry matter digestibility, are both lower. These nutrient contents are subject to less variation than with grasses and this particularly enhances their value as dry season feeds for livestock.

Protein

A significant variation in crude protein (CP) content occurs between species of trees and shrubs and even between edible parts of the same plant. Guérin reported a variation of CP content from 6 to 23% in dry matter (DM).

In general, leaves are higher in CP than twigs, almost twice as much in the case of southern African browse according to Walker. They also contain more CP on average than pods but the latter were found with higher organic matter and digestibility.

Leguminous species were found to contain 25 to 50% more crude protein than non-leguminous plants. Leaves of phyllodenous acacias tend to have higher crude fibre, lower crude protein, phosphorus, organic matter and digestibility than other acacias.

In Saudi Arabia, Mirreh and al Daraan found that perennial shrubs retained high CP content for a longer period than did annual shrubs. The CP in perennials varied from 14% in spring to 7% in winter while, in annual ones, a variation from 22% to 8% was recorded during the same spring season.

In addition to seasonal changes, environmental factors such as fire, substratum and amount of tree cover also influence nitrogen content. Barbero et al. reported a difference of 20% between the CP contents of an underwood calcicolous stand and an open silicicolous stand in a Mediterranean environment. Lay found that CP content in browse increased by 42.2% after burning. However, most of the benefits disappeared within a year or two.

Minerals and Vitamins

The concentration of calcium and potassium is usually higher than that of other minerals; the average being around 1 to 1.5% for southern African browse. However, abnormally high concentrations of some minerals such as sodium chloride and selenium may be found in browse.

Vitamin A, B, C and PP are present in varying concentrations in leaves, fruits and seeds of some trees and shrubs. Carotene content is usually high during early growth and declines rapidly with maturity, except in evergreen shrubs which tend to retain it for long periods.

Digestibility

Dry matter digestibility (DMD), which is related to nutrient composition, varied widely among tree and shrub species. Skarpe and Borgstrom, working in Botswana with Kalahari woody species reported a range in digestibility from 38 to 78%. Similar findings were reported by Wilson.

The digestibility of cellulose and cell walls decreased as the lignin to cellulose or lignin to acid detergent fibre ratio increased. The digestibility of crude protein did not always match the high CP content which characterises fodder trees and shrubs. For example Wilson found an apparent digestibility as low as 14% for Heterodendrum oleifolium containing 12.5% of CP while Atriplex vesicaria, also with 12.5% CP, had a nitrogen digestibility of 71.4%. Also there was not always a correlation between intake and digestibility; highly digestible stuff may be poorly consumed and vice versa.

Deleterious Substances

Tannins, mimosine, cyanoglucosides, fluoroacetate, oxalate, selenium, saponin and sodium chloride are among substances which have the potential of decreasing feed intake, animal productivity and even causing toxicity at high levels ingestion.

Tannins relate to a large number of phenolic compounds whose deleterious effects are not uniform. Some compounds like condensed tannins (CT) even have positive effects on nitrogen utilisation by protecting protein from microbial degradation. According to Barry and Manley, CT concentration of 20-40mg/kg in DM is ideal. Higher levels (76-90mg) were found to be detrimental to the health of animals.

Detannification techniques include heating, wilting, treatment with chemicals, storage of fresh leaves in polythene bags and bio-treatment using the fungus *Sprotrichum pulverulentum.*

Mimosine is found in Leucaena leucocephala. Its concentration, which varies with sub-species, is also affected by environmental factors such as soil and climate. Studies have shown that *Leucaena* can be used without harm as asupplementary feed for ruminants at a rate of 20-50% of the ration. Differences in minosine concentration could explain such variation in results.

Domestic ruminants, however, were seen to be better adapted to the consumption of trees and shrubs with high content of deleterious substances, provided there was a gradual ingestion of the browse. This phenomenon of adaptation, as well as the whole metabolism of antinutritional substances, still need further investigation.

Intake

Domestic ruminants utilise trees and shrubs both as browse in situ and "cut and carried" branches in the stall. On rangelands, animals have the advantage of selecting from a wide choice of browse and obtaining high quality feed. Consumption of various types of forage reduces chances of poisoning. However, the effective contribution of trees to the nutrition of livestock is not always realised due to inability to reach the browsable canopy of tall trees. The cut and carry system obviates the limitation of inaccessibility of the browse. The system also facilitates rational usage of trees and shrubs. Its disadvantages include damage to trees due to injudicious lopping and the imposition of a limited variety of feed which may increase the risk of poisoning.

Use of fodder trees and shrubs in stall feeding

The period of adaptation to new feed may be quite long, as shown in the studies of Le Houérou in Libya. The intake of 9 species of native and exotic shrubs offered either alone or in a mixture to ewes increased with time and stabilised after 3 to 5 months at levels 3 to 5 times the initial intake. The author also reported that consumption of mixed shrubs was higher than that of a single species. Wilting for 24 to 48 hours also increased the intake.

Many trials on the use of shrubs and leaves of trees to supplement either natural grasses or crop residues gained positive responses in livestock performance. Nitis reported a case where sheep and goats fed on *Pennisetum purpureum* supplemented with 0.3 to 1.8 kg *Gliricidia* per day gained 17- 27% more weight than the unsupplemented animals. With Bali cattle fed on 80% of natural grasses plus 9% of *Leucaena* and *Musa,* and 11% of tree leaves, the increase in weight was 58% more than that of the control group. ILCA studies in Nigeria established that the use of Leucaena and Gliricidia as supplementary feeds significantly increased the growth and survival rates of lamb. Statistical analyses of the data showed that each 100g of browse DM consumed per day raises the productivity index by 1.41 kg lamb weaned/ dam/year. Unfortunately, the interesting findings on supplementary feeding are not adopted by small-scale farmers, particularly in sub-Saharan Africa.

Exploitation of fodder trees and shrubs on rangeland

Owing to their dietary preference and the influence of animal size on the vertical utilisation of plants, there is little competition among domestic ruminants in the exploitation in situ of fodder trees and shrubs. According to Ibrahim, the proportion of browse eaten by cattle and sheep varies from 2-30% of their intake. With goats, the intake ranges from 25-50% in the rainy season and up to 75% or more during dry periods. However, as the above feeding behaviour study demonstrates, these estimates may vary widely.

Studies on the contribution of fodder trees and shrubs to the productivity of grazing animals are still limited. Reasons for this include difficulties encountered when implementing trials with paired areas and the negative effects of lopping or deforestation on the environment. A study by Rees on the evaluation of browse potential in cattle production is one of the few reports available. She used steers on 3 types of paddocks: natural, lopped and cleared.

Table 9 shows that lopped paddocks had the highest fodder potential for increased animal production, particularly when young steers (1-2 years) were used. Comparison of the performance of the 2-3 year old steers suggests that the efficiency of habitat exploitation and feed utilisation increased with age.

Table 9. Yearly live weight gain of 1-2 and 2-3 year old cattle on natural, lopped and cleared woodlands. 1969-71 Mbala Ranch.

Age of cattle	*Woodland*	*Lopped*	*Cleared*
1-2	13.7	90.8	40.6
2-3	46.6	100.0	97.5

Further monitoring of the effect of lopping showed that, within 3 years, most of the trees managed for browse either die or grow beyond the reach of animals. This study confirms that fodder trees and shrubs have potential for increased livestock production but has shown that the magnitude of the contribution to the productivity of grazing animals greatly depends upon accessibility of the browse.

High accessibility is not often achieved in the grazing situation prevailing in the dry tropics which carry the majority of domestic ruminants. Increasing accessibility of browse entails human interventions which may be detrimental to woody plants. There is therefore a need to develop appropriate woodland management systems which will ensure optimum utilisation concomitant with persistence of browse production in grazing systems.

8

Livestock and Environment

One of the great challenges facing the world over the next decades is to preserve its natural resources while at the same time producing sufficient food to satisfy the demands of a growing human population. World population is expected to grow from 5.5 billion now to about 8 billion in the year 2020. Incomes also continue to grow, especially in the developing world and future projections estimate an annual *per capita* income growth ranging from about 3 percent in sub-Saharan Africa and Latin America to about 6 percent in Asia. Furthermore, there is a strong population move from the rural to the urban areas, again primarily in the developing world. By the year 2000, approximately 44 percent of the world's population is expected to reside in urban areas, up from 30 percent in 1980. These trends will have immense consequences on the volume and composition of global food demand, especially in the developing world. Specialists of the International Food Policy Research Institute (IFPRI) estimate that the current demand of 1.7 billion tons of cereals and 206 million tons of meat, may rise by the year 2020 to 2.5 to 2.8 billion tons of cereals and at least 275 to 310 million tons of meat.

At the same time, alarming symptoms of deterioration of the resource base are being observed world-wide:

— *Land degradation*. Although there is substantial discussion on the extent of the problem, between 700 million and three billion hectares of land are reported degraded because of human activities.

— *Water scarcity and pollution*. Twenty-two countries suffer severe water scarcity and a further eighteen countries have dangerously low levels . In addition, much of the global fresh water supply is unsafe because of pathogens and industrial pollutants.

— *Global warming*. Global temperatures have risen by 0.3 °C to 0.6°C over the last century, together with a 26 percent increase in carbon dioxide and 115 percent rise in methane levels of the atmosphere. A further increase of the global temperature by 1.8°C is foreseen over the next 35 years.

— *Diminishing biodiversity*. About 160 bird and 100 mammalian species are known to have become extinct over the last three centuries and the rate of extinction is increasing. The rate of losses in other animal and plant species may be higher. McNeely *et al.*, estimates that more than 3,000 plant species and more than 500 animal species are in immediate danger of extinction.

The way livestock are kept and milk and meat is produced will be a key factor in the future health of the planet. Animal agriculture is one of the most important components of global agriculture and livestock is one of the main users of the natural resource base:

— livestock use 3.4 billion hectares of grazing land and the production from about one-quarter of the world's croplands. In total, livestock make use of more than two-thirds of the world's surface under agriculture, and one-third of the total global land area;

— livestock raising is the sole source of livelihood for at least 20 million pastoral families, and an important, often the main, source of income for at least 200 million smallholder farmer families in Asia, Africa and Latin America;

— livestock provide the power to cultivate at least 320 million hectares of land, or one-quarter of the total global cropped area. This would otherwise would have to be cultivated by hand tools resulting in harsh drudgery, especially for women, or by tractor power with an inevitable drain on foreign exchange;

— livestock provide the plant nutrients for large areas of cropland. For example, estimates carried out under this study showed that, for the tropical irrigated areas, manure provides nutrients of an estimated value of US$ 800 million per year;

— finally, livestock are an important asset for investment and insurance for hundreds of millions of rural poor, in situations where banks are often too remote and the banking systems too unreliable for safeguarding any savings a smallholder might accumulate.

Table 1. Regional growth estimates of demand for meat & cereals over the period 1990-2020.

Region	Meat	Cereal
World	60- 93	49-65
Developed	17-18	19-33
Developing	123-206	68-91
Sub-Saharan	141-194	136-161
Latin America	76-105	53-77
West Asia & North Africa	104-157	74-100
Rest of Asia	148-255	64-85

The importance of livestock production can be expected to increase over the next decades. While, in the industrial world, demand for meat and milk will probably plateau, or even decline, in the developing world, income growth and urbanization will fuel a strong increase in demand. Current levels of meat and milk consumption in the developing world are only about one-fifth of those in the industrial world. The surge in demand from present levels of 206 million tons to 275-310 million tons or more per year by 2020, will be especially strong in Asia and Africa where the demand for meat is expected to triple.

Implications for Natural Resources

The challenge will be to satisfy this substantial rise in demand for livestock products at a technological level which the natural resource base can sustain. The pressure will be heavy. Agriculture, including livestock production, already contributes, to a greater or lesser extent, to the degradation and erosion of these resources. Livestock production may specifically contribute to land degradation, the decline and pollution of water resources, the emission of greenhouse gases and the erosion of biodiversity. However, the negative role attributed to livestock is frequently a result of other pressures and distorted policies. These pressures and distortions should be addressed.

With good management, livestock production can also make a positive contribution to the natural resource base by enhancing soil quality, increasing plant and animal biodiversity and substituting for scarce, nonrenewable resources such as fossil fuels. Wherever possible these *win-win* (economically and environmentally attractive) scenarios should be promoted, and policies and technologies which do so should be identified. However, as the expected increase in demand referred to above is to be met, it is likely that negative effects of livestock production will continue to emerge. The challenge is thus to identify policies and technologies which mitigate any negative environmental impact but which, at the same time, satisfy the considerable demand for livestock products.

Framework for Livestock Raising

Livestock raising is carried out in many forms, usually referred to as production systems or agro-ecosystems. Production systems evolve as a result of agro-ecological potential, the relative availability of land, labour and capital and the demand for livestock products. Many production systems are currently at a sustainable equilibrium, with livestock being produced in harmony with nature and in environmentally sound systems. However, over the last decades, several production systems have lost this equilibrium because of the pressure caused by growing human populations and increased demand for animal products.

Based on the degree of integration with crops and its relation to land, the world's livestock sector has been classified in this study into three broad livestock production systems, i.e. grazing, mixed farming and industrial systems. For land-based production forms, i.e. grazing and mixed farming systems, a sub-division is then required to allow for differences caused by agro-ecological conditions and the ways in which livestock affect the natural resource base. The broad categories are:

— *Grazing systems*. These are systems based almost exclusively on livestock production, with little or no integration with crops. They are mainly based on native grassland. In terms of total production, grazing systems are of lesser importance because they supply only 9 percent of global meat production. Of this, three-quarters comes from Central and South America and of the Organisation of Economic Cooperation and Development (OECD). Livestock interact in these systems with land, water and plant and animal biodiversity, especially wildlife. Agro-ecological conditions strongly define the nature and scope of livestock-environment interactions in grazing systems. The study therefore distinguishes between arid, semiarid and sub humid, humid and temperate grazing systems. In principle, grazing systems are closed systems, where the waste product (manure) is used within the system and does not present a burden on the environment. Resource degradation, especially of land and biodiversity, is now developing in many of the world's grazing areas.

— *Mixed farming systems.* In mixed farming systems, crops and livestock production are integrated on the same farm. Globally, mixed farming systems produce the largest share of total meat (54 percent), and milk (90 percent). Regionally, the mixed farming systems of the OECD countries and Asia provide by far the largest share of these products, but also in sub-Saharan Africa, West Asia and North Africa (WANA) and Central and South America, mixed farming is the main system for smallholder farmers. Resource use in mixed farming is often highly self-reliant as nutrients and energy flow from crops to livestock and back. By definition, such a closed system offers positive incentives to compensate for environmental effects, making them less damaging or more beneficial to natural resource base. Because of the completely different approaches needed to address the environmental effects of mixed farming, this study distinguishes between mixed farming in the developing and in the industrial world. The main challenge is to identify those policies and technologies which allow these systems to grow while sustaining their environmental equilibrium.

— *Industrial systems.* These systems cover industrial types of production and small-scale urban or pert-urban production in developing countries. Both monogastric and ruminant production systems exist. They provide 37 percent of the total global meat production. These systems are open both in physical and economic terms. They depend on outside supplies of feed, energy and other inputs. These systems are strongly market driven, making them less resilient to market upheavals than other systems. Because of their many interfaces with the outside world, these systems, if not properly controlled, offer many opportunities to neglect their environmental costs. The challenge is to identify the regulations and incentives which force the polluter to internalize the environmental costs, at a minimum cost to the consumer.

— *Global overlays.* In addition to the site-specific and production system related impacts of livestock, such as land degradation in the arid zones, deforestation in the humid zones or livestock-wildlife interactions in the savannas, there are a number of effects which transcend the specificity of production systems. These are the global overlays, which include the environmental aspects of feed production, the emission of greenhouse gases, the erosion of wild and domestic genetic resources, and the management of waste. As in the industrial system, some of these impacts can be traced to one polluter (called "point source pollution"), and can therefore be controlled with appropriate regulation at the source. The other impacts cannot easily be traced to one polluter ("non-source pollution"), and the challenge is then to find the incentives to encourage all producers to reduce these emissions.

Importance of Different Production Systems

The rapid increase in demand for meat and milk over the next decades is likely to cause a dramatic shift in the relative importance of the main production systems in their contribution to global meat and milk supply:

— The growth potential for extensive grazing and roughage production is limited. In response to increased population pressure, good pasture land is being converted into cropland, leaving increasingly poorer land for grazing and mixed farming. Industrial production from pigs, poultry and beef is therefore likely to increase relative to production from grazing and mixed farming systems. Past trends and future projections clearly point in that direction. One relatively unknown factor in this discussion is the effect of global trade reform combined with a strongly growing demand for feed grains. The first projections of the effect of trade reforms show that cereal prices will increase more than milk and meat prices. This, combined with the increased demand for feed grains, will certainly lead to higher feed prices. Although this would seem to favour grass-based production, the limited growth potential of grazing and mixed farming systems, combined with significant potential for improvements in feed conversion efficiency, will probably result in continued strong growth of the industrial system;

— Pork and broiler meat production will increase, relative to beef production. This is a direct result of the better feed efficiency of pigs and poultry;

— Livestock production is therefore likely to become more crop-based. Furthermore, as it is generally less expensive to transport and store grain than the corresponding amount of meat, intensive livestock production will move nearer to urban centres;

— Growth will be mainly in Asia and Africa. In Asia growth will be very fast as meat demand almost triples till the year 2020;

— For various reasons, including higher energy prices and less stringent pollution controls, livestock production in the industrial world is likely to move to warmer areas;

— In the developing world, aided by the recent availability of better disease control technologies, there has been a strong shift in production from the arid zones and the highlands to the more humid zones.

These changes will have a major effect on the nature and degree of the livestock-environment interactions. Implications are:

— a shift from relatively closed systems, in which its waste products are being used within the system, to a more open system in which its waste products are not returned; and

— a move from systems with good opportunities for synergies and win-win situations, such as the conservation and use of indigenous livestock breeds, nutrient recycling and livestock-wild life integration, to systems where the opportunities for such win-win situations are much more limited.

Livestock Grazing Systems and Environmental Challenges

About 60 per cent of the world's pasture land, just less than half the world's usable surface is covered by grazing systems. Distributed between arid, semi arid and sub humid, humid, temperate and tropical highlands zones, this supports about 360 million cattle, and over 600 million sheep and goats, mostly in the arid rangelands.

Grazing systems supply about 9 percent of the world's production of beef and about 30 percent of the world's production of sheep and goat meat. For an estimated 100 million people in arid areas, and probably a similar number in other zones, grazing livestock is the only possible source of livelihood.

Grazing can be visualized as beautiful cows in lush pastures in north-western Europe or New Zealand-livestock in harmony with nature. Indeed, livestock can improve soil and vegetation cover and plant and animal biodiversity. By removing biomass, which otherwise might provide the fuel for bush fires, by controlling shrub growth and by dispersing seeds through their hoofs and manure, grazing animals can improve plant species composition. In addition, trampling can stimulate grass tillering, improve seed germination and break-up hard soil crusts.

However, many people associate grazing animals with overgrazing, soil degradation and deforestation. To them livestock keeping in arid regions of the tropics provokes images of clouds of dust, bleached cow skeletons and an advancing desert. The two most quoted sources are the Global Assessment of Soil Degradation, which estimates that 680 million hectares of rangeland have become degraded since 1945, and Dregne *et al.*, who argue that 73 percent of the world's 4.5 billion hectares of rangeland is moderately or severely degraded. In humid areas, livestock are associated with ranch encroachment and deforestation of tropical rainforests and competition with wildlife.

Prolonged heavy grazing undoubtedly contributes to the disappearance of palatable species and the subsequent dominance by other, less palatable, herbaceous plants or bushes. Such loss of plant and, in consequence, animal biodiversity can require a long regenerative cycle. Excessive livestock grazing also causes soil compaction and erosion, decreased soil fertility and water infiltration, and a loss in organic matter content and water storage capacity. On the other hand, total absence of grazing also reduces biodiversity because a thick canopy of shrubs and trees develops which intercepts light and moisture and results in overprotected plant communities which are susceptible to natural disasters.

The environmental challenge is thus to identify the policies, institutions and technologies which will enhance the positive and mitigate the negative effects of grazing. Environmental challenges, issues and options differ significantly according to climate and land capabilities. Livestock-environment interactions are therefore described separately for the arid, semi-arid and sub-humid, humid rainforest, and temperate and tropical highlands grazing systems respectively. As will be seen, that differentiation is particularly important for the arid eco-systems. As aridity increases, so does variability of rainfall, to the extent that the periodicity of rain becomes the single most important factor affecting the state of the natural resource base. Classical concepts of vegetation succession and climax vegetation do not apply in such environments and new concepts are required. These are detailed below.

Grazing Systems of Arid Areas

Probably more than any other system, at least until recently, arid rangelands have been associated with land degradation. The concept of "desertification" originated from the sight of degrading fringes of arid rangelands and advancing deserts. In addition, arid grazing systems and nomadic pastoralism have been associated with inefficient and backward production.

Over the last decade these views have changed radically. First, there is now significant evidence that in the arid zones the extent of land degradation is greatly exaggerated. Part of the doom view of continuing degradation is caused by problems of definition and time frame. Definitions were based on parameters, which could not easily be measured "diminution or destruction of biological material, which can lead to desert-like conditions", or mixed process and end state. Even more importantly, most assessments covered only a short period, and did not include the concept of reversibility, which seems critical from a development and long term sustainability perspective. If desertification is defined as an irreversible process of land degradation, there is less "desertification" than previously claimed.

Secondly, there is convincing evidence that traditional transhumance production systems on arid rangelands are highly efficient. Studies show that production of protein per hectare of traditional nomadic pastoralists in Mali and Botswana is two- or three-fold higher than production from sedentary production systems or ranching under similar climatic conditions in Australia and the USA respectively. In addition, arid grazing systems are often multiple-use systems, with wildlife and other plant products being important additional products.

The greatest degradation of land and vegetation is around settlements and water points. These areas are usually within a radius of about 1-5 km of the waterpoint and often look quite spectacular. However, assuming an average distance of 10 to 30 km between waterpoints, degraded areas would, even in the worst cases, amount to no more than 10 percent of the total area.

Biodiversity

Arid rangelands contain a broad variety of plant species for wild and domestic herbivores. For example, in Africa, there are perhaps as many as 3,500 plant species upon which the continent's herbivores feed, compared with less than 150 species on which humans depend. There is no widespread risk of an immediate irreversible loss of plant biodiversity. Mixed grazing systems help to maintain wide plant diversity and, because the vegetation is extremely resilient, any changes in the vegetation are likely to be the result of an unusually dry period and, therefore, temporary.

Driving forces

The way most dryland ecosystems are traditionally used explains, to a large extent, their resilience. Arid rangelands have traditionally been used under a communal property regime by nomadic producers who move their stock in search of pasture according to season. From the wet season grazing they will move their animals to higher-potential river valleys, cropland or mountain meadows (the "key resources") for the dry season. With highly variable rainfall (both in time and space), pastoral economies are typically of the "bust and boom" type: a "boom" when rainfall is plentiful and herds and flocks grow, and a "bust" when drought occurs and animals die. Thus, abiotic factors such as rainfall, rather than livestock density, determine long term primary production and vegetation cover.

A substantial body of evidence from the last decade, shows that arid regions contain dynamic and highly resilient ecosystems, with a strong capacity to regenerate rapidly when the rains return. Similarly, traditional pastoral systems have conserved biodiversity because pastoralists have a direct interest in preserving a wide variety of plants and animals. Gathering range products, such as medicinal plants, gums and resins, is an important part of the pastoral way of life.

The fundamental driving force on natural resources is population pressure, especially that applied from outside the arid rangelands and their traditional inhabitants. While population growth of pastoral peoples has been slow, growth of non-pastoral groups in the arid and semi-arid regions have been among the highest in the world. This strong growth of other groups results in an increasing encroachment by arable farmers on to the pastoralists' "key resource" sites.

Flood plains which have been traditionally used for wet season grazing, such as the Interior Delta of the Niger and the Senegal Valley in West Africa, and also smaller sites of high potential, are being converted into cropland. Furthermore, flexibility of animal movement is progressively hampered by increased population pressure and loss of corridors between wet and dry season grazing areas. Stock is increasingly concentrated the entire year on the same lands, breaking the ecologically sound cycle of alternating use of wet and dry season grazing areas, leading to over-use of dry season grazing land and, inevitably, to human suffering. Such increased pressure often results in war, as shown by the recent confrontation between Senegal and Mauritania and the many armed conflicts in East Africa.

Increased population pressure also leads to greater water development and permanent human settlement in arid rangelands. Although the direct effect of waterpoints on land degradation is relatively limited, the development of water supplies for more intensive use can upset an entire eco-system. It may lead to de facto privatization of land around the water point and change the relationship between traditional wet and dry season grazing areas, changing traditional dry season grazing into year around grazing.

But not all forces are exogenous to the pastoral system. In the Borana region of southern Ethiopia, because of indigenous population growth, the number of milk producing cattle per head of the pastoral population is falling. In consequence, the pastoralists face increasing poverty, decreasing food energy levels and greater risk from the effects of drought. Furthermore, the larger number of people in the region means that fuelwood is being cut at an ever greater rate. The inevitable result of all these pressures is land degradation. Drought often exacerbates such situations and, indeed, the ability to recover after drought is one of the main indicators of long term environmental and social sustainability of arid grazing systems.

Several human activities exacerbate the fundamental driving force of increasing population pressure. They are:

Stabilization of the system. Often well-intentioned policies sought and still seek to stabilize the "boom and bust" cycles which exist between man, animals and vegetation in arid rangelands. Examples of these are:

— *Attempts to regulate the stocking rate.* This was the main focus of many rangeland development projects in the seventies and early eighties. Development projects promoted individual ranch development, or the allocation of grazing areas to groups, under strict maximum stocking rate obligations. Such attempts failed completely because of the irrelevance of any carrying capacity estimate under variable conditions and because of the difficulty of enforcement. In addition, these attempts reduced the essential mobility and flexibility of the system, and concentrated stock on limited grazing areas, which may have received little rain.

— *Feed subsidies for drought relief.* Under the argument that it would be necessary to protect national livestock assets, North African and Middle Eastern governments, in particular, have had a policy of subsidizing feed grains as emergency assistance in times of drought. Besides the questionable social effect and the doubtful overall economic benefit, subsidized feed also has a harmful effect on the rangelands. The tendency is for too many animals to be retained on the rangeland, thereby preventing normal regeneration of the vegetation after drought.

Changes in access to land. Traditionally, Muslim land tenure rules favour crop farmers over pastoral herders in access to land because "laying of the hand"confers exclusive ownership. This was exacerbated by nationalization of arid rangelands which was introduced by many governments in the postcolonial period in Africa and Asia and undermined still further the intricate fabric of customary practice. An ecologically well-balanced system of communal land use degenerated into a "free for all" open access system in much of sub-Saharan Africa and India or was replaced by individual farms, such as in Algeria, Botswana and Zimbabwe.

Individual farms were too small to permit an efficient use of the erratic rainfall patterns inherent in these areas. In communal areas, the traditional collective internal discipline in the management of the resources disappeared and overgrazing and land degradation followed. Jodha demonstrated this most convincingly in India. The same mistakes are now occurring in the Middle East and are likely in Central Asia. Under the slogan "privatization is desirable", state farms are privatized, the more productive land falls into the hands of the most powerful individuals and sound management of the entire resource base is put in jeopardy.

Inappropriate incentives. Under a nationalistic strategy of food production self-sufficiency, many governments, have subsidized tractors and fuel and supported high producer prices for domestic cereal production, further encouraging the encroachment of crops in the "key resource" sites of the arid zones. Many of these areas are marginal for crop cultivation and would not have been cultivated without the subsidies. However, as noted before, these key resource areas constitute the "safety net" for rangeland livestock production and are the key to environmentally sound arid land management. Cheap fuel and trucks have also led to "motorized nomadism" which has allowed the wealthier herders to move their animals rapidly to wherever pasture was available. This has also interfered with the normal regeneration process of the emerging vegetation. Finally, high meat import tariffs also contributed to over stocking.

Societal response

The over-riding need is to stop the building-up of further

human pressure in arid zones. Arid rangelands are already efficiently used and no significant increases in productivity can be envisaged. Since populations in these areas experience a downward spiral of declining livestock/people ratios and increased vulnerability to drought, employment generation outside the dry rangelands is therefore the most critical component of developing sustainable rangeland resource use in the arid zones.

As the second priority, although still of crucial importance, external interventions in the system need to take account of the non-equilibrium status of pastoral systems in arid zones and enhance rather than restrict flexibility and mobility. This means that attempts to regulate stocking rate should be stopped. First, the carrying capacity of rangelands in these non-equilibrium environments cannot be estimated with any acceptable degree of reliability. Second, such estimates can even be dangerous, as they may lead to the wrong intervention in an attempt to control stocking rates. Third, even apart from the technical flaws in the estimation of the carrying capacity, experience has shown that it is almost always impossible to enforce stocking rates.

The third priority should be the strengthening of traditional pastoral institutions and resource management practices. "Getting the institutions right", by empowering pastoral people, is now generally considered the main challenge in pastoral and arid land development. Furthermore there has been a growing acknowledgment of the multiple uses that arid rangelands provide to a wide variety of users.

Specific actions to support these strategy elements are to develop effective co-management regimes, forging partnerships between the State and a wide variety of users, with the State carrying the overall responsibility for arbitrating conflicting interests at national level, and facilitating negotiation. Practical management decisions and negotiations between competing users should, however, be delegated to the local level. The delegation of powers would aim to strengthen:

— the application of the "subsidiarity" principle, within the local administrative capability which requires devolution to local groups of those public administrative powers that undermine or duplicate traditional governance structures. The creation of pastoral associations has already been an important focus of arid land institution building, although results have been mixed;

— customary resource user rights. The need for flexibility and mobility would require the boundaries to be "fuzzy", although with more clearly defined territorial boundaries for the "key resources";

— conflict resolution mechanisms, with clear sets of procedural tasks for formal law, and clearer definition of when customary law takes precedence.

The fourth priority is the identification of effective drought management policies. Irreversible land degradation in arid zones, if it occurs, originates as a result of high stocking rates during droughts. The appropriate drought strategy is to de-stock as early and rapidly as possible, rather than seeking to maintain maximum stock numbers. Policies and subsequent investments which support such rapid de-stocking are the development of savings and credit schemes, and infrastructure investments in roads, markets, slaughterhouses and cold storage. However, little experience has been gained of the application of these concepts in dis-equilibrium systems. The only notable exception is Australia.

Incentive policies also play an important role in arriving at sustainable rangeland use. Key elements of appropriate incentive policy would be:

— Increasing the costs of grazing in order to reduce animal pressure would promote earlier off-take. The perception of pastoral people keeping livestock for wealth and social status has been abandoned and there is now a general acceptance that pastoralists' behaviour is, as may be expected, economically rational. This economic rationale should induce herders to forsake liveweight gains of their almost mature and only slowly growing animals, and sell them earlier, because the cost of the grazing would outweigh the benefits of the final liveweight gain. But cost recovery has to be combined with a decentralized management of the proceeds by the beneficiaries, to provide appropriate feedback mechanisms to producers and could cover:

— Levying grazing fees for communal areas. Such fees have been proposed often but little experience exists of them in practice. A system of progressive fees, with larger herds paying more per head, is attractive, particularly for many of the dry areas in Africa where livestock are increasingly concentrated in the hands of outside owners, such as civil servants, traders and crop farmers;

— Full cost recovery especially for water supply and animal health services. In many cases, water has been a free resource supplied by the public sector. Full cost recovery, including construction costs, should most likely reduce the number of large boreholes and, therefore, reduce local degradation around these water points;

— Removal of price distortions for other agricultural inputs, in order to reduce the conversion of pastoral key resources into marginal crop land.

Most donor agencies now accept these as basic elements of their pastoral strategy as shown, for example, by the excellent recent review by IFAD of common property resources. In addition, African and Middle Eastern government authorities are increasingly adopting participatory and decentralized approaches, with a greater degree of cost recovery, although considerable apprehension still persists in allowing pastoral groups any great degree of independence.

The key areas where future research is urgently needed emerge from the above recommendations. With limited opportunities for increasing production, the emphasis must be on conservation. This implies:

— the identification of appropriate indicators to provide reliable information on resource trends in arid areas;

— the development of appropriate methodologies for economic appraisal of investment in the conversion of pastoralists' "key resources". These appraisals normally compare only the pasture yield during the dry season or summer grazing, with the cereal yields and, in most cases, the economic benefits from the increased cereal yields outweigh the pastoral use. If the impact of crop encroachment on the entire eco-system is calculated, the use of such high potential areas for livestock is economically more efficient, and certainly environmentally more efficient, than conversion into cropland;

— continued research on pastoral institutions;

— the design of sustainable drought preparedness plans, with particular emphasis on decentralized management and the design of appropriate banking and insurance schemes; and

— the identification of appropriate conflict resolution schemes.

Grazing Systems in Semi-arid and Sub-humid Tropics

Although these two zones are combined in this discussion because of their similar physical livestock-environment interactions, many of their agro-ecological and socio-economic conditions are different. The semiarid zones are generally densely populated. Typical areas include parts of the Sahel, the rainfed crop-livestock areas of North Africa, some of the rangelands of Central Asia and the drier areas of the Indian subcontinent. The sub-humid zone savannas have, until recently, been rather sparsely populated because access was impeded by human diseases and tick-borne diseases in animals. Typical areas include the savannas of West Africa, eastern Colombia and the Cerrados of Brazil, southern and eastern Africa and eastern areas of the Indian subcontinent.

These areas are now becoming the main frontier for agricultural development and worldwide it is where ruminant livestock numbers still grow, as they accommodate, in sub-Saharan Africa an overflow of animals from the drier areas, and, in South America, animals from the Andean Highlands. Livestock interaction involves most components of the environment e.g. land, water and biodiversity. More reliable and more abundant rainfall allow higher population density, more cropping, and more perennial grass and shrub species but also implies fewer possibilities for herd flexibility and mobility. As a result, pressure on land resources mounts and land degradation becomes more severe. Although data on a global basis are not available, data from a transect in Mali for example, showed that land degradation in the 600-800 mm rainfall areas was significantly greater than in the 350-450 mm rainfall area. In the higher rainfall area, the percentage of bare soil increased from 0 to 10 percent over the period 1950 to 1990, whereas in the more arid areas there was no significant change.

Irreversible degradation is also reported in the semi-arid regions of North Africa and the Middle East. The amount of available fodder from rangeland per hectare declined, indicating long term decline in range productivity. For the former Soviet Union, Gilmanov, estimated that about one-third of the total area of 370 million hectares of rangeland was degraded although, from his figures, no decline in fodder availability per hectare can be detected. Zonn estimates that for the Kalmyk Republic in Southern Russia 6 million hectares out of a total of 7.6 million is moderately to heavily degraded. Irreversible land degradation in the sub-humid savannas of Africa and South America is still minor.

With a high population density, the development of livestock and human drinking water supplies has been a key aspect of rangeland development in semi-arid zones where surface water is a constraint. Proportionally, degraded lands extend over a larger area in the semi-arid zones than in the arid zones, as a result of higher population densities and greater concentration of water points. As in arid zones, a circular area of degradation of 1-5 km radius around the water point is quite normal, although these "sacrifice areas" are normally quite fertile and are surrounded by bushes at the outer edge of the sacrifice area which balance out the negative effects. Long term studies in Senegal and Sudan did not find a significant expansion of the sacrifice areas of individual water points in these countries.

As the environment becomes more humid, the animal disease threat, especially that imposed by parasites, also increases. In Africa, in particular, control of livestock disease "vectors" such as ticks and tsetse flies is therefore essential for livestock development. Ticks have traditionally been controlled by cattle dips or sprays using organo-chlorines. Several donor-supported projects promoted this technology in the seventies and early eighties. Inappropriate dosing resulted in increased tick resistance, and inappropriate drainage of dip liquid caused water pollution and damage to the fauna. Very little quantitative data are available on the impact of these acaricides on flora and fauna but it can be assumed to be substantial at the local level.

The semi-arid and sub-humid tropics are one of the world's most important repositories of plant and animal biodiversity. Wildlife is an important income earner for African countries, with a large part of the wildlife population outside the protected areas. For example, in Kenya, 70 percent of the large mammals are outside the game parks, and many are under considerable stress. Ottichilo reported a 30 percent decline in these populations over the last 20 years.

In the sub-humid savannas, weed invasion is a major problem threatening biodiversity, although the role of livestock is only secondary here. For example, the grass *Imperata cylindrica* in the Philippines and Indonesia has now infested more than 5 million hectares. Invasion with broad-leafed plants and shrubs is more common in the savannas of Africa and the Americas.

Driving forces

Population growth and poverty are the underlying factors determining land degradation and biodiversity loss in the semi-arid zones. The human population in the semi-arid countries of sub-Saharan Africa has grown at a rate of 3.2 percent over the last decade, resulting in greatly increased demand for food and fuel. Many countries in the semi-arid zone belong to the poorest in the world. Land degradation is thus clearly a result of cropping in fragile areas, fuelwood collection and overgrazing by the prevailing sedentary livestock production systems. Several attempts have been made to determine the relative share of each of these pressures.

The UNEP study estimated that overgrazing causes 35 percent of land degradation whereas agricultural activities and deforestation is responsible for 28 percent and 37 percent respectively. For Africa, the estimate is that overgrazing contributes to 49 percent of the total degraded land area. For the USA, tourism and military bases are also important causes of rangeland degradation.

However, with multiple and interlocking forces affecting rangeland degradation, such precise allocation of the causes of degradation seems an oversimplification. It might even be dangerous because it can easily lead to the wrong type of intervention. Nevertheless, it is important to understand the dominant pressure factor in any region in order to be able to identify the correct mitigating measures. For example, quite clearly, to lessen land degradation caused by fuelwood cutting requires different measures than if overgrazing is the main cause of degradation.

A better approach for defining the relative weight of the different causes of land degradation is to define the sustainable human population for each factor. Steeds and Gore did this for different West African eco-zones. For the Sahel, fuelwood is the most limiting factor in meeting the subsistence requirements of its human population. For the higher rainfall areas, livestock production remains below that necessary to satisfy subsistence requirements.

For all the zones, however, fulfilling food subsistence needs with a combination of livestock and crop production exerts less pressure on the land than satisfying fuelwood requirements.

In the sub-humid savannas, increased cropping leads directly to growth in the livestock population. For example, Bourne and Wint found that, for every 10 percent increase in cropping, the livestock numbers grew by about three TLU per km^2. This is a beneficial strategy, mainly brought about by livestock herders' need for markets and crop farmers' need for organic fertilizer.

However, heavy livestock grazing in open park land savannas leads to the disappearance of the most palatable grass species where they will be replaced by other herbaceous plants. This leads to increased encroachment of shrubs and bushes which, in the normal vegetation succession, leads to forest regeneration. The forest would not develop into a closed canopy but would, in this system, where bushfire is a normal phenomena, become an open parkland. This cycle usually takes about 30 years. Overgrazing in savannas thus leads to initial bush encroachment, with increased biomass production, but, because it can no longer be grazed, the diversity of plant and animal species declines. In some cases the bush encroachment leads to the use of economically and environmentally expensive herbicides.

The interaction between wildlife and livestock in these eco-systems is complex. First, there is increasing evidence of a possible complementarily, and only limited competition, of wildlife and livestock in grazing. As already shown by Schwartz and Ellis the grazing "overlap" between most wildlife species and livestock is limited. Mwangi and Zulberti and Western showed that the combination of livestock raising and wildlife management resulted in an equal or better species wealth than either of these activities individually. Furthermore, in national parks in Kenya, such as Amboseli, where livestock are not permitted, biodiversity is decreasing, and there is a corresponding increase in unpalatable species and bush-encroachment. On the other hand, as the same author points out, there are many degraded areas in Kenya due to combined wildlife-livestock pressure. The picture is thus not uniform but potential for benign and even synergistic co-existence certainly exists and needs to be encouraged. About 70 percent of big game animals are outside national parks and, because population pressure is likely to preclude a large increase in the area or number of such parks, the preferred approach must be to maintain a better wildlife habitat outside the park system.

The driving forces leading to losses in animal biodiversity are habitat destruction, species introduction and hunting. In an analysis of the threats to mammals in Australia and the Americas, UNEP argues that pastoral development affects about 30 percent of the species, compared to logging, which affects about 25 percent, and cultivation which affects 55 percent of mammalian species on those continents. Habitat destruction is playing an important role in the developing world, especially in the sub-humid savannas.

Road construction, and immigration from the drier areas, leads, in Africa, to the destruction of the habitat of the carriers of African sleeping sickness. In turn, this removes the protection of wildlife which is tolerant to the disease. International agencies, including GTZ and the World Bank, have financed extensive tsetse clearance campaigns in West Africa. Traditionally, these campaigns used a combination of hand sprayers and aerial spraying of insecticides, initially with organo-chlorines, to eradicate the tsetse fly.

Nagel argues that pesticides from this period are still found in some African birds. Since the mid eighties, compounds with shorter residual effect, such as synthetic pyrethroids have been used. These second generation compounds caused substantial initial damage to flora and fauna but permanent effects were not observed with single spraying. For example, in the World Bank funded tsetse clearance project on the Adamaoua Plateau of Cameroon, where permanent damage and high residue levels were found, they appeared to be the result of repeated spraying in the border areas.

Bush encroachment, as a result of inappropriate grazing management was reported to be the most serious environmental damage. Land use plans have often been advocated as an essential pre-condition for tsetse clearance, and international donors have made the preparation of such plans conditional to the financing of the eradication campaigns. However, experience with the enforcement of such land use plans has been dismal because local authorities lacked the authority and means for their enforcement.

Traditional land tenure practices have tended to be less robust in these more humid areas than in the drier areas. This lack of strong traditional tenure practices has been particularly significant in sub-Saharan Africa where after tsetse clearance operations, uncontrolled settlement patterns developed.

In addition, hunting and culling of wildlife was encouraged in the past, because wildlife in general was considered to be a reservoir of diseases, such as Rinderpest and Malignant Catarrhal Fever and vectors of disease, such as East Coast Fever and Trypanosomiasis. However, the control of the above mentioned diseases has improved considerably and there is a much better understanding of which species harbour which specific diseases.

Policy pressures

Inappropriate policies, whether institutional or those governing incentives, often exacerbate the negative effects of livestock on land and biodiversity and undermine some of the potentially positive effects.

The effect of price policies has often been to discourage livestock off-take and thereby increase the pressure on grazing:

— *The urban bias.* Many African governments maintained overvalued exchange rates and allowed the importation of highly subsidized milk and meat from the EU in order to provide cheap food to urban populations. For the semi-arid and sub-humid grazing systems, especially in West Africa, this dumping of meat and milk reduced off-take and increased grazing pressure. The recent devaluation of the CFA franc almost doubled off-take from the Sahelian herds, thus reducing grazing pressure, and almost eliminated the imports of dumped meat from the EU;

— *Absence of banking institutions.* Similarly, with high inflation rates yielding a low return on deposits and, in many cases, the absence of any form of reliable banking system, livestock have become the main form of investment in many developing countries. This is the case not only for the pastoralists themselves but also for urban investors, who rely on hired labour for herding their animals. Although difficult to quantify, some estimates suggest that as much as 50 percent of the livestock in the Sahelian countries are now owned by outsiders. These outside investors have different objectives, less environmental concern and less respect for traditional tenure rights;

— Under investment and market failure in plant and animal biodiversity. Even where there is no deliberate attempt to distort markets, valuable natural habitats are converted to agricultural production, including livestock, because other costs, such as loss of biodiversity, have not been accounted for. Such adverse conversion would be less likely if an appropriate value, including future potential value, was given to the natural vegetation, including medicinal plants and animal wildlife.

The value, costs and returns from wildlife, compared to livestock and agricultural production, is highly variable. At the level of the national economy, the opportunity cost of wildlife biodiversity conservation in protected areas in terms of foregone livestock and agricultural production seems to outweigh the income from tourism and forestry generated by these protected areas.

Inappropriate institutions constrain sustainable wildlife livestock integration on communal areas. Traditionally there has been a rigid, centralized and regulatory attitude of public institutions in the protection of wild animals. This was especially the case in East Africa, where wildlife management was typically organized by central administrations in a rather military fashion.

There was no benefit sharing with the local population, even though wildlife brings high financial cost to the local population because of crop damage, livestock loss to predators and diseases transmitted from wildlife. This has lead to the antagonistic reactions of many herding and farming communities. In addition, the prohibition in many countries of sport hunting and consumption denies important potential benefits from wildlife.

Societal response

The prevailing combination of poverty and high population growth, which characterizes the countries of the semi-arid regions of the world, cannot easily be broken. This will mean continued pressure on the remaining trees and shrubs for fuel, on higher potential land for crop production, and therefore more pressure on the grazing resources. Nevertheless, livestock populations can be expected to grow, especially in the sub-humid areas. Physically, these areas are characterized by much more stable climatic conditions, and traditional grazing management concepts of stocking rate control and individual land tenure become more valid as the precipitation increases.

The introduction of market pricing for inputs and outputs will promote more sustainable livestock production. On the institutional side, again, most of the arid land recommendations concerning decentralization and local empowerment apply also here.

Community-based wildlife management is now generally accepted as an important component of sustainable wildlife management in communal areas. Key issues involve:

- — the strength and cohesion of the traditional social fabric, as shown by the difference between the successful community management around the Maasai Mara park and the difficulties experienced by a similar initiative in Amboseli where traditional authority is much weaker; and
- — achieving an equitable distribution among the local population of the proceeds of wildlife revenue.

For the African savannas, little progress has been made in developing more appropriate land tenure systems, and land use planning is still as ineffective as in the past.

Furthermore, many of the areas cleared in the 1980s have been re-invaded by the tsetse fly. Most donors have stopped funding tsetse eradication for this reason and, where tsetse eradication still continues, such as in the EU funded programme in southern Africa, this is done with environmentally more benign methods and a stronger grassroots participation. The impact of those programmes still has to be assessed. In any case, the threat of the tsetse fly is diminishing, as population pressure destroys the habitat of the fly, with or without eradication campaigns. Finally, especially for these areas, there is a great need for alternative investment possibilities, such as savings banks, to divert savings away from investment in livestock.

Before the introduction of technologies intended to rehabilitate degraded land, it is essential that institutional arrangements are in place. For the semiarid rangelands, once the institutional framework is in place, it should be possible to:

— develop water supplies. Before any development is undertaken, very careful consideration should be given to the possible impact on the distribution of wet and dry season grazing, to settlement patterns and on the efficiency of feed utilization. There is convincing evidence that increased watering frequency increases the metabolic rate and hence decreases the efficiency of the utilization of scarce feed resources;

— regenerate the vegetation, through resting, over seeding or planting of adapted fodder and fuelwood shrubs. The latter has been widely practiced in North Africa although with limited success;

— breed for hardiness in indigenous breeds, with selection schemes based on producers' herds;

— introduce multi-species, to encourage a better utilization of the vegetation. There is growing evidence that the combination of different species, including wildlife and livestock has a positive effect on plant and animal biodiversity;

— introduce overall "management" approaches, such as Holistic Resource Management, which establishes clear goals, including environmental ones, and then selects the most appropriate tools for particular site conditions. This approach is strongly based on the principle, that it is not the total grazing pressure, but the continued grazing on the same area, which causes degradation. Timing of the grazing is therefore of critical importance.

— improve infrastructure, especially roads and markets, to allow a smooth movement of goods and services and rapid herd off-take, especially in times of drought.

For the sub-humid tropical savannas, the emphasis has to be on the development of multiple uses, combining sustainable farming systems with integrated wildlife-livestock management and the establishment of a few protected areas for key plant and animal species. However, with increasing population pressure, this becomes more difficult. Furthermore, many government agencies in the developing world are too weak and, in the developed world, too dependent on political fashion, to rely exclusively on protected areas for biodiversity conservation, especially for big game of the sub-humid savannas. Eco-tourism and commercial hunting would need to be integrated with commercial livestock production and mixed farming.

There is a wide array of technologies available:

- for the management of the natural savannas, regulated burning, i.e., deliberately using fires at different times of the year to affect different classes of plants, can be one of the most efficient ways of maintaining a good tree/shrub/grass mixture. However, regulating savanna burning has not worked in the past because of weak enforcement agencies, and this has led to a general distrust, and often a blanket prohibition, of burning by departments responsible for rangeland and savanna management. This technology may not work if governance is weak unless there is strong participation by the local population;
- for vector control, new and environmentally more benign methods are now available. For tick control, vaccines are available for the diseases spread by a number of sub-species, although not yet for East Coast Fever. Long duration tick sprays, which can be applied on individual animals (pour-on) now exist, and these have a negligible environmental effect. A number of environmentally benign control methods also exist for tsetse control; and
- CIAT has developed a means of improving savanna in South America with improved grass and legume species in rotation with rice, and ILRI has proposed improved fodder production using leguminous plants.

Grazing Systems and Tropical Rainforests

Tropical rainforests cover about 750 million hectares, with half of that area located in four countries: Brazil, Zaire, Indonesia and Peru. These rainforests are estimated to contain 50 percent of the worlds' plant and animal species. For this system, the interaction of livestock production and biodiversity is therefore the main focus in this study although there are also significant effects on land and air quality. Ranch encroachment in tropical rainforest has probably, more than anything else, typified the negative effects of livestock development and, while generally the effects have been negative, even here positive examples exist.

The transformation from tropical forest to crop and pasture land brings about substantial losses of soil fertility and soil erosion. Furthermore, in many tropical rainforest areas, pastures can only be sustained for a short period of up to ten years. Soil nutrients are rapidly depleted after clearing and grasses are soon replaced by less useful vegetation. Natural regeneration of forests is quite difficult, especially in large degraded areas. More than 50 percent of the pasture areas in Amazonia have now been abandoned in a degraded state.

For the rainforests, data on biodiversity losses are dramatic. Since 1950, about 200 million hectares of tropical forest have been lost, with the result that of some unique plant and animal species, in one of the world's richest sources of biodiversity have become extinct. Forest areas of Central America have declined from 29 to 19 million hectares, since 1950, although, since 1990, the rate of deforestation in this region has fallen. In Central America in the 1980s rainforests disappeared at the rate of 430 thousand hectares per year but this declined to 320 thousand hectares over the period 1990 to 1994. In South America, the deforestation rate in the 1980s was about 750 thousand hectares per year. It is not known, whether this rate has declined over the last years.

Much of the deforested areas in Latin America went into ranching, sometimes after initially being cropped. In Central America, pasture areas have increased from 3.5

million to 9.5 million hectares, and cattle populations have more than doubled, from 4.2 million head in 1950 to 9.6 million in 1992. In Brazil, about 70 percent of the deforested areas are converted into ranching. In Asia and sub-Saharan Africa, the decline in the forest area is mainly the result of crop expansion and not livestock related.

Driving forces

Livestock ranching has often been regarded as the driving force behind the deforestation of tropical rainforests. However, there is increasing evidence that livestock ranching in deforested areas is merely the most obvious symptom of a much more complex degradation process with a variety of driving forces. They are detailed below.

Road construction leads to accelerated deforestation, and is possibly the single most important direct cause of deforestation. For example, Ledec found that in Panama, for each kilometre of an all-weather road, between 400 and 1,000 hectares of forest were cleared. Roadside ranchland was sold in Nicaragua at about three times the price of comparable land one day's walk away. In Rondonia, Brazil, the paving of an existing road was the main factor in clearing 100,000 km2 of rain forest, and focused, in the early eighties, the world's attention on this relationship. Livestock ranching, with little supervision requirement and few bulky inputs became an attractive activity along these new roads.

Arable farming, in the rainforest areas, usually takes the form of "slash and burn" agriculture, especially in the forest areas of sub-Saharan Africa and Asia. Several studies describe this as the most important mechanism of deforestation in these areas. Overall the area cropped increased over the last two decades. When population pressure is low, and the area deforested is relatively small, secondary forest climax vegetation can return, thus limiting longer term soil erosion. If population pressure increases further, a downward spiral of declining soil fertility and crop yields emerges. In Central and South America, the total area under crops has remained stable, but slash and burn cultivation continues at the primary forest frontier, and is compensated by desertion of earlier depleted cropping areas, or their conversion into pasture. This conversion of crop land to pasture is common practice, as many ranchers use a crop of corn to generate income for future ranching.

Relative importance of different pressure factors in deforestation. As in the case of desertification, there have been several quantitative estimates on the individual importance of these factors. However, unlike in the case of desertification, the direct cause is easier to ascertain, making these estimates more relevant. Nevertheless, there are some overlaps, especially as logging is often a precursor for cropping, which in turn often precedes ranching. This means that even in tropical rainforests the cause of environmental damage is hard to apportion.

Policy pressures. Probably more than for any other livestock- environment interaction, inappropriate incentives, land tenure and institutional policies have played a major role in deforestation. land degradation and erosion of biodiversity in the humid tropical areas.

Inappropriate incentives from governments and international agencies in support of livestock development, played an important role in the destruction of the Latin American rainforests. Especially in the seventies, the livestock sector secured a disproportionate share6 of credit at subsidized rates and with lenient re-imbursement conditions and control. Most of these loans were invested in land or deforestation, and in turn raised land prices even further. However, it appears that the role of subsidized credit is limited. For example, Ledec showed that only 7 to 10 percent of the deforestation in Panama could be attributed to subsidized credit. A similar picture emerges from Brazil. The large ranches greatly benefited from investment subsidies, tax holidays, and subsidized interest rates. However, the large ranches caused only a small part of the deforestation and most ranching in Amazonia is in the hands of medium to small size ranchers, who have had less easy access to these subsidies.

Beef exports to the USA, frequently quoted as one of the main causes for ranch development in Latin America, were important in Central America in the 1960s and early 1970s, when international beef prices were high. But the Amazon area never produced more than 5 percent of the total beef supply of Brazil, and that production was not exported. Furthermore, declining global prices in the eighties, and increased protection in the USA market, especially following NAFTA, have reduced exports from the other Latin American countries. However the fall in prices seems to have had only a limited effect on deforestation. It may have pushed traditional livestock ranches into cropland, but it has not halted expansion in the frontier zones. Ranchers have had very few alternative opportunities.

Regarding land tenure, the key phenomenon was the land grab in rainforests, which occurred as a result of a general increase in land prices in Central and South America over the last decades. "Money laundering, labour remittances and a hedge against inflation" kept prices above the productive value. The almost permanent price increase made land a safe investment and led to increased speculation. Finally, recent government settlement programmes and the establishment of internationally funded protected reservation areas, raised prices further. A large part of the expansion of pasture land may therefore have had more to do with land speculation than with cattle raising per se. Deforestation for ranching thus became more a titling strategy than one based on an economic activity.

Institutional requirements concerning land titling procedures reinforced these trends. Several Latin American countries actually prescribed occupation, and thus deforestation, as a condition for giving out ownership titles. As clearing land for ranching was much easier than establishing proper cropland, ranching became the preferred occupation to fulfill the title requirements. Furthermore, land clearing discouraged squatters. For example Edelman reported that practically all 13 squatter invasions that occurred in one particular area of Costa Rica, took place in heavily forested areas, and not in ranching country. Ranchland was psychologically better protected. Finally, some governments even encouraged deforestation for livestock production by colonization schemes.

Technology and policy options

Many causes for ranching encroachment in rainforest areas have disappeared. Subsidized credit for ranching has been phased out almost everywhere and the overall volume of credit to the sector has declined. In addition, some of the more exogenous factors turned against ranch expansion. Meat exports from Central and South America have declined dramatically and world market prices for beef have dropped over the last decade. All these measures reduced investments by large and urban investors in ranching, and brought down the overall rate of deforestation. However, these changed incentives had little effect on the small and medium producer, and deforestation for crops and small-holder ranching continues. A new look at the policy instruments for these farmers is thus required, as additional action is necessary to halt the still on-going conversion of rainforest into ranches. The following measures have been suggested:

— *Discourage road construction.* Road construction is the single most important factor in deforestation. Discouraging road construction would probably be the most powerful deterrent for ranch and farm establishment and even though it may be politically difficult, this should seriously be considered for highly valuable forest areas;

— *Establish protected areas, and expand the rights and protection of indigenous land use rights.* While effective protection of large areas may be difficult to implement under the prevailing situation of understaffed, underpaid and poorly motivated government staff, the most valuable areas can be set aside. Testing of revenue-sharing mechanisms of the proceeds of rainforests, involving all stakeholders could be tested, to improve the incentives for forest conservation;

— *Introduce taxation for pasture and crop land to discourage conversion of forest to crop and pasture land.* This is an attractive concept although, with poor or even non-existing property registry service in many of the rainforest areas, it will be very difficult to enforce;

— *Promote more research and extension.* Kaimonitz especially stresses these needs, as he sees slow technological change as one of the key factors favouring extensive livestock production. Indeed, the basic ranching system has changed very little since the early 1960s, and no agricultural alternatives to ranching have been developed.

Grazing Systems in Temperate Zones

Grazing systems in the temperate zones cover about 440 million hectares, or 13 percent of the world's pasture lands. They contain about 2 percent of the world's cattle and 10 percent of the small ruminants. They are mainly based on natural permanent pastures, and occur mostly in China, northern USA and South America and the CIS. They contribute about 5 percent of the global beef supply, 12 percent of mutton supply and 3 percent of the global milk supply.

Most grazing areas in the temperate zones are in relatively good vegetative condition. They were originally formed by the grazing of wild animals and, because livestock grazing is not fundamentally different, there is no convincing evidence that the change from wild to domestic grazing has, in general, had any long term detrimental effect. Long term stocking trends both in the USA and Australia show an initial high stocking rate in the late 19th century and early 20th century, falling off after a period of about ten years, to a lower level which is then sustained over the following decades. This indicates a fairly stable eco-system. Indeed, independent observers argue that rangelands in the western United States are now in a better ecological condition than at any other time in this century. For example, in 1987 there was half as much "poor" rangeland but double the area of "good" rangeland as in 1964. In this equilibrium environment, there is ample experimental evidence that there is no difference in erosion and water infiltration between light, moderately grazed and un-grazed areas. In addition, there are some excellent examples of how grazing with domestic stock can improve biodiversity and how grazing can reduce fires and promote seedling establishment by reducing biomass accumulation.

However, there are regions where the vegetative condition of temperate grazing land is less satisfactory, notably in the CIS and Asia, western Europe and the western United States. Grasslands in the CIS are often heavily overgrazed. This is caused by the significant preference given under the previous command economy to industrial, mechanized production, with a strong dependence on feed grains. Meadows and grazing animals did not fit well in such systems and therefore received scant attention. Although incentives have changed, there is still a dearth of appropriate technology to improve pasture lands in the CIS and overgrazing, resulting in increased erosion and reduced water infiltration, is still a common phenomenon. Overgrazing in the temperate, high altitude pastures of Central Asia is particularly severe after the transition to a market economy.

Many grazing areas of western Europe and the eastern USA are over fertilized. The highly subsidized and guaranteed milk price under the Common Agricultural Policy (CAP) of the EU in the seventies and eighties encouraged strong intensification, and this led to very high nitrogen fertilizer use on pastures. For example, in the eighties, the average nitrogen application on grassland in the Netherlands was about 500 Kg N per hectare per year, whereas the average N outflow in milk and meat did not amount to more than 82 kg N per hectare per year. As a result, these meadow soils are now overloaded with nutrients, especially nitrogen and phosphorus, contaminating ground and surface water.

In the western USA, there is currently strong concern about the impact of grazing along the banks of rivers and streams: the riparian areas. Such stream sides, which normally comprise less than 5 percent of the total area, receive often 20-30 percent of the use and therefore water quality, plant biodiversity, and trout yields may be impaired. It is estimated that more than 50 percent of riparian areas are seriously damaged. On the other hand, there is ample evidence that nutrient leakage into streams is minimal where streamside pastures remain in good condition.

Driving forces

Policies continue to drive livestock production into degrading temperate grazing systems:

— In Central Asia, changes in fuel prices and privatization of land are currently the driving forces causing overgrazing. Previously, annual movements of animals were greatly facilitated by low energy prices but market pricing of transport has reduced animal movements and is changing the traditional cycle of seasonal grazing and rest periods into year-round grazing;

— In western Europe and the USA, the most significant driving forces to have an impact on the environment are those policies which foster intensification of livestock production. The Common Agricultural Policy (CAP) of the EU (then European Economic Community), which heavily subsidized dairy production encouraged intensive production at high stocking levels and therefore excessive use of fertilizer and concentrate;

— The phasing out of feed subsidies in Central Europe is still too recent to have had a measurable effect;

— Grazing on public lands in the western USA is being driven by a larger concern over how public lands should be utilized. Currently riparian areas are the focus of the debate. In the past, the importance of riparian areas was under estimated

> and they were largely viewed as sacrifice areas, while upland grazing was the focus of range improvements. Also contributing to the stress on riparian areas has been the inappropriate conversion of sheep permits to cattle permits. One particularly important area of this debate has been the cost charged for grazing public lands. Some have argued that the grazing land is under-valued and that fees are therefore too low.

Technology and policy options

The first priority in reducing the environmental impact of the over-fertilized grazing systems in the EU and USA, and enhancing their environmental condition, is to phase out the current system of milk and meat subsidies. The current shift in EU policies to replace the subsidies on these products by subsidies on farmers' income, and tie this to landscape maintenance, is a very positive development. In addition, much can be done to achieve a better nutrient management. Work in the Netherlands shows clearly the possibility of reducing input levels by introducing a ley farming system with more farm grown fodders, careful application techniques and lower nitrogen inputs, without substantial reduction of the output. Experience from Europe and the USA shows that farmer education is a key factor in arriving at better nutrient management. The introduction of a mineral balance at farm level has helped farmers to recognize the excess nitrates and ammonia which are emitted. Experience with farmer study groups in the Netherlands, clearly shows that it is possible to decrease the nutrient loading by 25 to 50 percent with an improved financial result on many farms. Similar results are emerging from the USA, where, under the motto "it is better that we do it now voluntarily, than have it imposed by the

Government", farmers, for example, in the Chesapeake Bay area, are drastically reducing their fertilizer input levels.

A greater reliance on market mechanisms, combined with a less restricted use of wildlife, are essential elements in the management of riparian areas in the western USA. Once such land is available to the highest bidder and exploitation of wildlife is attractive, more land would go into wildlife or combined livestock-wildlife production, and grazing pressure would be more evenly spread. Some regulation of the riparian areas might be necessary and community-based landscape objectives and values would probably be more effective.

9

Improving the Safety of Livestock Products

The contribution of livestock to food supplies in developing countries is increasing at a higher rate than that of cereals. The total meat production in developing countries rose from 30 million tonnes in 1970 to 69 million tonnes in 1990, and it is projected to reach 105 million tonnes in the year 2000 and 143 million tonnes in 2010.

Developing countries now account for almost half of the total world meat production. While in developed countries a certain saturation and decrease in the overall consumption of livestock products can be seen, the demand for livestock products will continue to rise in developing countries as a result of their rapidly expanding populations and the tendency towards higher incomes in most of these countries.

In contrast to developed countries, where the high consumer intake of livestock products has contributed to health problems, the predicted increase in consumption patterns in developing countries over the next few years will be beneficial to the nutritional and health status of the majority of their populations. Together with fish, livestock food products provide essential amino acids, minerals and vitamins in a concentrated form, and their fat content can also supply much needed calories. Meat is particulary important in the diets of young children and pregnant women because of its high protein and iron content.

In spite of the increasing levels of livestock production in most developing countries, the proportion of meat in the diet of the average consumer remains rather low. This is mainly because the human population grows almost as fast as that of livestock, but also because meat is scarce in many places and its cost is comparatively high.

The supply shortage of livestock products in certain areas in developing countries is a consequence of the traditional marketing systems, where fresh, unprocessed meat is sold at meat markets a few hours after slaughter. These systems are used in all the developing regions where no cold chain for refrigeration exists.

Although livestock is raised in rural areas, meat is frequently not available there, since the slaughtering takes place in central abattoirs in urban areas. Cost also limits the consumption of livestock products. In many developing countries, the average daily income is equivalent to the price of 1 kg of meat. It is therefore not surprising that many consumers can only occasionally afford livestock products and they are consequently not part of the everyday diet.

The average annual per caput consumption of meat is 20 kg, compared with 80 kg in developed countries. The latter is calculated on the basis of carcasses and edible offal. In the poorest countries, the annual rate of meat consumption per caput is as low as 5 kg. In actual fact, the consumption patterns of most populations are probably even less, since the average figures also include the higher income groups with atypically high consumption. It is therefore clear that a greater consumption of livestock products would be highly desirable in developing countries.

The following examples illustrate the benefits of a higher intake of livestock products in developing countries:

— For adequate nutrition, a person's daily protein intake should be 1 g per kilogram of body weight. In developed countries this is easily reached and, in most cases, exceeded, while in developing countries there is a widespread deficit. Ideally, 30 to 50 percent of the daily protein intake should be animal protein, particularly in developing countries, since this provides an optimal range of amino acids. The average intake of animal protein in developing countries, however, is as low as 15 g per person per day, compared with almost 60 g in developed countries.

— With regard to minerals, meat is one of the most important sources of iron. In developing countries anaemia caused by lack of iron is one of the major nutritional problems, particularly in young women and children. The other important livestock food, milk, is the most important source of calcium in human nutrition. It can also be used as an additive in processed meat products.

With regard to providing consumers with adequate quantities of quality meat and meat products, one aspect is becoming increasingly important, namely, the processing of products based on meat as a raw material. Appropriate processing techniques offer the opportunity of overcoming the two main constraints to a better supply of livestock products - availability and price.

Shelf-stability

Using specific processing techniques, it is possible to produce shelf-stable meat products that can be stored under ambient temperatures, eliminating the limitations of the traditional livestock-products marketing system. In the traditional system, pressure is put on producers and consumers to complete the cycle from procurement to consumption of livestock products in a very short period, in most cases, within one day. In contrast, in prolonged distribution chains for livestock products, which will be increasingly evident in the rapidly expanding population centres in all developing regions, not only can shelf-stable livestock products be handled safely, but their supply to consumers in rural areas far from food processing plants and abattoirs will be ensured.

Affordability

Meat processing can also tackle the price problems of meat products, while at the same time allowing the raw materials, such as carcass meat and by-products, to be classified as valuable and less-valuable products. The parts that are graded "less valuable" because of poorer censorial properties such as taste and tenderness, may be very valuable from the nutritional and processing point of view. This is the case with meat derived from the head, lower parts of the legs, the diaphragm and internal organs. These parts can be used for processed meat products. It is obvious, therefore, that good-quality products can be produced with cheaper raw materials, while the more expensive carcass parts can be profitably sold to customers who can afford higher-priced meat or meat products. Prices for meat products can be further decreased by using meat extenders, which may include plant or crop products such as flour and starches. Dairy products and fish, which can partially replace expensive meat while keeping the products at an adequate level of nutritional quality, may also be used.

Meat Preservation and Processing Techniques

Tables 1 and 2 list meat preservation and meat processing methods that are already practiced to a certain extent in some developing regions or that will be of increasing importance in order to supply the expanding populations with sufficient quantities of good-quality and affordable meat products.

Table 1 Meat preservation methods

Method	*Basic technologies*	*Main effects*
Lowering the moisture content	Evaporation of product water by drying or replacement of water by other food additives	Control of microbial growth
Increasing the acidity	Lactic acid fermentation or addition of organic acids	Inhibition of growth of spoilage and food-poisoning microorganisms
Preservative effects of treatment or meat additives	Salting, curing or smoking; suitable preservative	Inhibition of growth of spoilage and food-poisoning microorganisms
Thermal treatment	Boiling, cooking or roasting	Significant reduction of most micro-organisms, but recontamination is unavoidable as product is not hermetically sealed
Thermal treatment in hermetically sealed containers	Pasteurisation, sterilisation	Reduction of bacterial counts and enzyme activities by heat; no recontamination

In many cases, the preservation and processing technologies mentioned are not applied as a single treatment, but rather one treatment method may follow another or a combination of treatments may be used.

Intermediate Moisture Meats

The evaporation of water or the replacement of part of the meat with non-meat additives can often help to extend the shelf-life of a product. The substitution of meat with plant products or other meat extenders decreases the water content of the meat products, or, expressed in scientific terms, the water activity in the product. The product's microbiological stability increases when less water is present for microbiological growth.

Table 2. Meat processing methods

Method	*Products*
Carcass dressing at the abattoir level	Dressed whole carcasses, carcass sides or quarters Edible animal by-products
Carcass/meat cutting	Wholesale and retail meat cuts for frying, cooking, roasting, etc.
Drying of unsalted pieces of meat	Dried meat for preparation of meat dishes
Salting and curing of meat cuts with or without heat treatment	Cooked ham, cooked cured beef, bacon, raw ham and other raw cured products such as dried cured products
Coarse comminuting of raw meat	Hamburger-type products
Coarse or fine comminuting of raw meat with subsequent fermentation	Fermented sausages of the salami type, fermented Asian products
Fine comminuting of raw meat with subsequent heat treatment	Emulsion-type sat/sages, e.g. frankfurters, meatballs, emulsion type meat pies
Fine comminuting of precooked meat together with internal organs, blood, etc.	Liver paté, liver sausage, blood sat/sage, meat pies.

The technical term for food products with reduced water activity is intermediate moisture (IM) foods. All empirical meat preservation methods based on the drying of the meat or meat products fall into this category. Drying can either be done alone or in combination with other methods such as salting, curing or smoking. Meat drying techniques have been used for centuries and can be considered the oldest meat preservation methods.

Meat drying is simply the exposure of strips or flat pieces of fresh meat to the sun or open air. This treatment reduces the meat's water content so rapidly that no bacterial spoilage can take place, even though the temperatures remain high. There are certain disadvantages with this simple drying method, however, including contamination by insects, dust and other environmental impacts. Cabinet solar dryers have been especially developed for this purpose, where warm air is conducted through a hermetically closed chamber, thereby avoiding any undesirable impacts from the outside. So far no source of energy other than the sun has been used in solar drying methods.

Solar tunnel drying is a more sophisticated approach, where high energy inputs from the sun are achieved through a long solar collector using electric fans. In areas without electricity, these fans can also be run by using photovoltaic panels to convert solar energy directly into electricity.

The construction materials for the solar tunnel dryer are relatively cheap and simple and can be produced locally. This drying method is expected to be a major step forward towards hygienically, technically and economically efficient meat drying, also on a semi-industrial scale if the daily throughput is high enough. The method is absolutely clean and has no negative impact on the environment.

Of course, this improvement of the drying methods must be accompanied by improvements in the pretreatment of fresh meat and in the packaging and storage of dried meat. Different pre-treatment methods, such as selecting specific carcass parts, salting, curing and smoking, will be tested, as will traditional and modern packaging materials such as vacuum bags. The greatest variety of dried meat products can be found in Africa. Besides the common dried meat strips or flat pieces, there are many different product preparations and names. In Latin America, particularly in Brazil, a product called *charqui* has long been used to supply remote areas. Several large flat pieces of beef are salted, piled on top of each other and finally exposed to the sun for drying.

A North American product is pemmican, or jerky, which is dried beef or game meat, often kept under a layer of lard for better storage. In Asia, simple meat drying, although known, is not very popular. Particularly in Southeast Asia, meat to be dried is usually pretreated, for example, it may be soaked in a salt or sugar solution.

Sugar plays an important role in meat preservation in Southeast Asia, especially for those products that have not been dried. Sugar replaces part of the product's water and makes it more stable from the microbiological point of view. Moreover, sweet-tasting meat products are very popular there. A well-known example is Chinese sausage, which contains a mixture of minced pork, fat, salt, spices and sugar filled into small casings. Stored under ambient temperatures without further treatment, the sausages are shelf-stable for a few days. Heat treatment by cooking or frying only takes place immediately prior to their consumption.

Fermented Products

Other types of shelf-stable meat products are fermented

Hygienic Slaughtering

With regard to further processing of meat, one important aspect that is widely neglected in developing countries must be stressed. Both modern and traditional meat processing methods require that raw meat of good hygienic quality is supplied to the meat processors through abattoirs. Unfortunately the abattoir sector does not keep apace with new developments in the meat processing sector.

Meat processing methods is usually a private enterprise, while abattoirs are often publicly run or public facilities leased to private entrepreneurs. The recent change to market economies in many countries has only exacerbated the problem, since a great number of small hygienically unsatisfactory slaughtering facilities, mostly run by private businesses, have opened up.

It has also been observed that reasonably good central abattoirs have been abandoned and cities are now being supplied by many clandestine slaughterhouses because of lower costs resulting from the absence of expensive hygienic facilities and proper meat inspection. This development is clearly detrimental to any progress in meat processing. It is often the case that meat processors cannot obtain raw meat of good hygienic quality for further processing, a fact that also has serious consequences for the quality of processed products.

It is therefore imperative that urgent measures are taken in the abattoir sector. If the situation remains unchanged, serious constraints in supplying expanding populations with meat and meat products can be expected.

In 1987, FAO launched a programme for the establishment of small-scale modular slaughterhouses in rural areas in order to improve hygienic slaughtering and meat handling and to reduce meat losses. Since adequately trained staff are essential for the proper functioning of slaughterhouses, workers who are to operate the small-scale modular slaughterhouses are trained under this programme.

Meat Inspection

Another issue often neglected in developing countries is that of adequate meat inspection. The principal reason for meat inspection is to protect consumers from zoonotic diseases. It has also been introduced in order to condemn those carcass parts that are heavily contaminated or have undergone unfavourable biochemical changes and are therefore not suitable for further processing.

Training in Meat Processing and Sanitary Control

Training is one of the important issues in meat technology, facilitating the introduction of proper meat processing and meat preservation methods. In Asian countries, regional courses are held periodically to teach participants traditional as well as modern meat processing and preservation methods. Of utmost importance is the hands-on training provided to the participants to enable them to perform and disseminate the techniques in their own countries. Similar activities have been carried out in Latin America and more are planned for the coming years.

One of the main constraints in the improvement of hygiene and the technical quality of slaughtering, meat handling and by-product utilisation is the lack of adequately trained personnel at all levels. In addition to those related to the establishment of small-scale modular slaughterhouses, regional meat training courses have been conducted in Botswana for a period of six years and courses of a shorter duration have been held in the Sudan, Thailand and Costa Rica. Similar courses are foreseen in the near future.

Prospects of Meat Processing

A recent review commissioned by FAO on agroprocessing predicts a staggering increase in the consumption of livestock products in developing countries. Experience shows that this will be associated with an increasing proportion of processed meat products in the overall meat consumption. There are two reasons for this:

- With increasing meat consumption, consumers are keen to have a greater variety of meat preparations at their disposal, a demand that can be met by offering various types of processed meat.
- Greater availability of carcass meat increases the amount of "manufacturing" meat, that is, less valuable meat parts not suitable for meat dishes, but which can be fully utilised for processed products. The accessibility of sufficient raw material will stimulate meat processing, which is an optimal and very economical way of using all available carcass parts for human consumption.

In the near future, meat processing technologies and equipment will be introduced into most developing regions. In every region it can be observed that the different levels of development of individual countries are reflected in the technical standards of their meat processing capacities. In some countries, at least in their major cities, Western standards have almost been reached, with attractive processing and sales facilities for a variety of meat and meat products, all kept under refrigeration. Other countries still have less sophisticated facilities or products, but they are certain to catch up in a few years with even the most advanced countries. There is also a third category of countries that have very traditional meat marketing methods, where a longer period of structural and technical development is required.

Bibliography

Barrett, J., "Cultivating Community: Local Business People and Family Farmers Sharing Values and Mutual Support", *Maine Organic Farmers and Gardeners Association*, Unity, Maine, 2003.

Berton, V., "50 Ways to Sustainable Farming: Diversification in Nebraska: Raising food, not feed, raises profits", *Featured farmer: Tom Larson, Field Notes*, NC SARE Quarterly Fact Sheet; Winter/Spring 2001.

Cramer, C., "Sustainable Farming Connection: Where farmers find and share information", Web site, *Committee for Sustainable Farm Publishing*, 1997.

David and Kay James, "Farming and Ranching for Profit, Stewardship & Community", *SARE 2000 conference proceedings*, SARE, 2001.

Davis, W., "A profitable, sustainable ranch", In: McDermott, M., ed., *Future Farms: New Ideas for Family Farms and Rural Communities. Conference Proceedings*, February 8 and 9, 2000.

Digiacomo, G. et al., "Sustainable Farming Systems: Demonstrating Environmental and Economic Performance", *Minnesota Institute for Sustainable Agriculture* (MISA); University of Minnesota, Saint Paul, MN, 2001.

Hall, B. and Kuepper, G., "Making the Transition to Sustainable Farming", ATTRA (Appropriate Technology Transfer for Rural Areas), Fayetteville, Ark., Dec. 1997.

Joannides, J., Kivirist, L., and Ivanko, J., "Farming and ranching: Livestock", *Renewing The Countryside*, Web site. 2005.

Mulla, D., Everett, L., and DiGiacomo, G., "Whole Farm Planning: Combining Family, Profit, and Environment", Minnesota Institute for Sustainable Agriculture; University of Minnesota Extension Service , St. Paul, MN, 1998.

Salatin, J., "Balance: Stability for your life and farm", Acres USA, *A Voice for Eco-Agriculture;* 32(4):1-3,. 2002.

_______________., *Family friendly faming,* Acres USA, A Voice for Eco-Agriculture; 30(9):1-5, 2000.

Savory, A., "The New agriculture. The Agriculture Vision 2000 Conference", *Sustaining the Agricultural Community in the New Millennium,* Great Bend, Kansas, January 11, 2000.

Sullivan, P., "Holistic Management: A Whole-Farm Decision Making Framework", ATTRA (Appropriate Technology Transfer for Rural Areas), Fayetteville, Ark., July 2001.

Van Der Pol, J., "Conversations with the land: The relationship between livestock and the land", *Greenbook 2000, Energy and Sustainable Agriculture Program,* Minnesota Department of Agriculture, St. Paul, Minn., pp. 4-6, 2000.

Index